# THE THEORY OF LARGE-SCALE OCEAN CIRCULATION

Mounting evidence that human activities are substantially modifying the Earth's climate brings a new imperative to the study of ocean circulation. This textbook provides a concise but comprehensive introduction to the theory of large-scale ocean circulation as it is currently understood and established. Students and instructors will benefit from the carefully chosen chapter-by-chapter exercises at the end of the book. This advanced textbook is invaluable for graduate students and researchers in the fields of oceanic, atmospheric, and climate sciences and other geophysical scientists as well as physicists and mathematicians with a quantitative interest in the planetary fluid environment.

R. M. SAMELSON is a professor of oceanic and atmospheric sciences at Oregon State University. He received a BS in physics from Stanford University and an MS in mathematics and PhD in physical oceanography from Oregon State University. He has served as editor of the American Meteorological Society's *Journal of Physical Oceanography* and on numerous scientific program and review committees. He has authored or coauthored 85 peer-reviewed scientific publications on topics in physical oceanography, atmospheric science, and geophysical fluid dynamics as well as a book on Lagrangian motion in geophysical flows and several book chapters on various aspects of physical oceanography and geophysical fluid dynamics. This book is based on lecture notes that he has developed over the past decade for a graduate course in large-scale ocean circulation theory at Oregon State University.

# THE THEORY OF
# LARGE-SCALE OCEAN CIRCULATION

R. M. SAMELSON

*Oregon State University*

CAMBRIDGE
UNIVERSITY PRESS

# CAMBRIDGE
## UNIVERSITY PRESS

University Printing House, Cambridge CB2 8BS, United Kingdom

One Liberty Plaza, 20th Floor, New York, NY 10006, USA

477 Williamstown Road, Port Melbourne, VIC 3207, Australia

4843/24, 2nd Floor, Ansari Road, Daryaganj, Delhi - 110002, India

79 Anson Road, #06-04/06, Singapore 079906

Cambridge University Press is part of the University of Cambridge.

It furthers the University's mission by disseminating knowledge in the pursuit of education, learning and research at the highest international levels of excellence.

www.cambridge.org
Information on this title: www.cambridge.org/9781108446709

First published 2011
First paperback edition 2017

*A catalogue record for this publication is available from the British Library*

*Library of Congress Cataloging in Publication data*
Samelson, R. M. (Roger M.)
The theory of large-scale ocean circulation / R. M. Samelson.
p. cm.
Includes bibliographical references and index.
ISBN 978-1-107-00188-6 (hardback)
1. Ocean circulation. I. Title.
GC228.5.S26 2011
551.46′2–dc22 2011001980

ISBN 978-1-107-00188-6 Hardback
ISBN 978-1-108-44670-9 Paperback

The water sparkles
secret messages of light
I want to learn the code
   —N. Samelson, *Home to the Mockingbird* (1971)

# Contents

# Preface

The purpose of this text is to give a concise but comprehensive introduction to the basic elements of the theory of large-scale ocean circulation as it is currently understood and established. The intended audience is graduate students and researchers in the fields of oceanic, atmospheric, and climate sciences and other geophysical scientists, physicists, and mathematicians with a quantitative interest in the planetary fluid environment.

When I first began to study the physics of ocean circulation, it was the intrinsic scientific interest of the subject that was most apparent and appealing to me. Since that time, evidence has grown strong that human activities are substantially modifying the Earth's climate, with long-term effects that threaten to significantly disrupt the environmental structures on which human life and civilization depend. This troubling development brings a new imperative to the study of the ocean's large-scale circulation as this circulation and its interactions with the atmosphere and cryosphere play a clearly important, but still poorly understood, role in the global climate system. Although the ocean components of most numerical climate models are based on the primitive equations, the dynamics that they represent are essentially those of the planetary geostrophic equations described here, because of the necessarily coarse horizontal resolution of climate-model computational grids. Thus, the present material should be of particular interest to climate dynamicists.

The text is based on lecture notes that accumulated over roughly the last decade, during which I regularly taught a core graduate physical oceanography course on the theory of large-scale ocean circulation. I am grateful to the students in the past few years of these courses who have responded favorably to earlier drafts of these notes and encouraged me to complete them into a text. That these notes did accumulate is, to me, the main argument for finishing them into a text: their existence proves that though much excellent material is available elsewhere, the precise trajectory through the material that has appealed to me differs measurably from other treatments.

Accordingly, it should also be clear that what the text presents as established understanding is, of course, a personal view.

My hope has been to keep the presentation as straightforward as possible. Accordingly, the plan of the book is essentially linear: it is meant to be worked through from start to finish, with not too much in the way of optional topics or branching logic. Some material beyond what I have normally covered in lectures has been added, so it may not be possible to get through the complete text in a single course, especially if alternative topics or perspectives are also to be included. The first six chapters cover topics and models that may be considered classical and generally, if not universally, accepted. Chapters 7 and 8 steer a course toward less charted waters, presenting perspectives that are less well established and based on more recent research in large-scale physical oceanography. Chapter 9 touches on the thermohaline problem, which is otherwise largely neglected in favor of a development using a single density variable, and Chapter 10 contains some brief closing remarks. The emphasis throughout is on the derivation and exploration of the basic elements of the theory rather than on comparison with oceanographic measurements. The equations are, for the most part, left in dimensional form; the figures, however, are mostly in dimensionless form. The reader is encouraged to work through the associated transformations between dimensional and dimensionless variables. Several exercises are provided for each chapter; these range from the straightforward computation and plotting of numerical results for solutions derived explicitly in the text to the use of independent analytical reasoning to obtain extended, related results.

The majority of the material presented in this text is the accumulated result of the efforts of many individual scientists over many decades and longer and is not the result of my own research. As is customary for introductory pedagogical texts, citations of original publications have generally not been added so as to not clutter the narrative line. A few are included in the notes sections that close each chapter to give the reader some entry points into the research literature. Closely related general references, portions of which have substantial overlap with segments of this text, include the excellent texts by Huang (2009), Pedlosky (1987, 1996), Salmon (1998), and Vallis (2006). Some of the exercises are motivated by specific publications such as Huang and Pedlosky (1999), Ledwell et al. (1998), Rhines and Young (1982b), and Webb (1993). I would also like to acknowledge here the general and multi-layered debt of gratitude that I owe to so many members of the extraordinarily generous and dedicated ocean research community, to which I have had the privilege to belong for longer now than I care to admit in print.

Some portions of the text, especially in Section 5.6 and Chapters 7 and 8, closely follow material originally published by the author (or the author and a coauthor, for Section 9.5) in the American Meteorological Society (AMS) *Journal of Physical Oceanography*, on which the AMS holds the copyright. The associated figures are reprinted by permission of the AMS, as indicated in the corresponding captions;

similarly, the redrawn Figure 5.8 is used with permission of the *Journal of Marine Research*. Ocean observations used in the figures were obtained from the *World Ocean Atlas 2005*, which is publicly available from the National Oceanographic Data Center of the U.S. National Oceanographic and Atmospheric Administration, and were processed using routines from MATLAB Oceans Toolbox, which is publicly available from the Woods Hole branch of the U.S. Geological Survey.

I am grateful to Matt Lloyd of Cambridge University Press for his support, encouragement, and patience throughout the project. Many authors have found that book manuscripts eventually can take on a life of their own; when I found that this one had stowed itself away in my backpack for a late fall hike up the Whitewater Trail to Jefferson Park in the Oregon Cascades, I knew it was time to turn it loose, with the hope that a few readers will find in it an approximate match to their own tastes. If the result is ultimately to improve our general community understanding of the physics of large-scale ocean circulation, then the effort will, at least in my own mind, have justified itself.

*R.M.S.*
*Corvallis, Oregon*

# 1

# Basic Physical Principles and Equations

## 1.1 The Large-Scale Ocean Circulation

Systematic observations of the fluid properties of Earth's ocean, made primarily over the course of the last hundred years, reveal coherent features with scales comparable to those of the ocean basins themselves. These include such structures as the subtropical main thermoclines and anticyclonic gyres, which appear in all five midlatitude ocean basins, and the meridional overturning circulations that support the exchange of waters across the full meridional extent of the ocean, from the polar or subpolar latitudes of one hemisphere to the opposing high latitudes of the other. These features and motions, which prove to be connected by robust dynamical balances, constitute the large-scale circulation of the Earth's ocean.

The global field of long-term mean sea-surface temperature is dominated by the meridional gradients between the warm equatorial regions and the cold poles but contains significant zonal gradients as well (Figure 1.1). The global long-term mean sea-surface salinity field has a more complex structure (Figure 1.2), with isolated maxima in the evaporative centers of the midlatitude subtropical gyres. The global sea-surface density field computed from long-term mean temperature and salinity reflects the competing influences of temperature and salinity on density (Figure 1.3). These fields are the surface expressions of complex three-dimensional interior property fields. The downward penetration of the warm equatorial temperatures is generally limited to the upper one-fifth of the water column (Figure 1.4), while salinity perturbations are less strongly confined to shallow depths (Figure 1.5). Lateral density gradients, more important for the large-scale circulation than vertical density gradients, largely reflect temperature variations in the upper subtropics and salinity variations at high latitudes and great depths (Figure 1.6). Coherent, large-scale structures in these property fields include the mean downward slope toward the west of isothermal and isohaline surfaces in the subtropics and the associated eastward zonal gradients of density (Figure 1.7). These distributions support and are in part maintained by coherent,

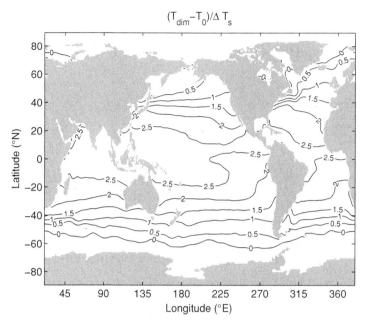

Figure 1.1. Dimensionless long-term mean sea-surface temperature $T = (T_{\text{dim}} - T_0)/\Delta T_s$ vs. longitude (°E) and latitude (°N) for the world ocean, where $T_{\text{dim}}$ is the observed temperature (K), $T_0 = 275$ K, and $\Delta T_s = 10$ K.

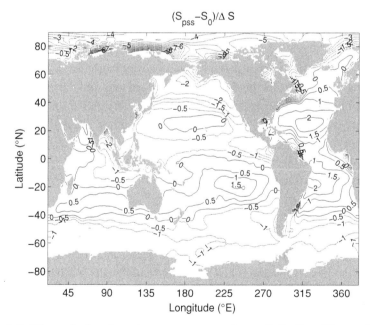

Figure 1.2. Dimensionless long-term mean sea-surface salinity $S = (S_{pss} - S_0)/\Delta S$ vs. longitude (°E) and latitude (°N) for the world ocean, where $S_{pss}$ is the observed salinity, $S_0 = 35$, and $\Delta S = 1$ for salinity on the practical salinity scale.

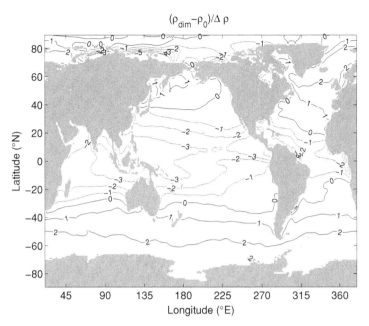

Figure 1.3. Dimensionless long-term mean sea-surface density $\rho = (\rho_{\text{dim}} - \rho_0)/\Delta\rho$ vs. longitude (°E) and latitude (°N) for the world ocean, where $\rho_{\text{dim}}$ is the observed density (kg m$^{-3}$), $\rho_0 = 1025$ kg m$^{-3}$, and $\Delta\rho = 1$ kg m$^{-3}$.

large-scale fluid motions, which in turn form one component of the spectrum of motions that is observed to occur throughout the world ocean.

The broad range of spatial and temporal scales over which ocean fluid motion occurs, together with the intrinsic nonlinearities of the general fluid dynamical equations, makes the direct analytical or numerical solution of complete fluid models of ocean circulation utterly intractable. However, theoretical progress can be made by using observed knowledge of the general characteristics and scales of observed features and motions to guide appropriate simplification of these general equations. For the large-scale circulation, a basic set of equations—the planetary geostrophic equations—can be deduced from a relatively limited set of a priori assumptions regarding the scales of the motion. From a theoretical point of view, this approach is essentially deductive in nature: it results in a system of equations for the purely large-scale motion that does not depend on assumptions about, or models of, the mean effect of smaller-scale motions on the large scale. In their simplest form, these planetary geostrophic equations are equivalent to a single partial-differential evolution equation that is first order in time but higher order in space.

Unfortunately, for the theoretician, this deductive model is not closed: it is not possible to pose a well-defined general boundary or initial-boundary value problem for these equations without introducing additional assumptions regarding small-scale motion near boundaries. The theory of large-scale ocean circulation thus rests on a

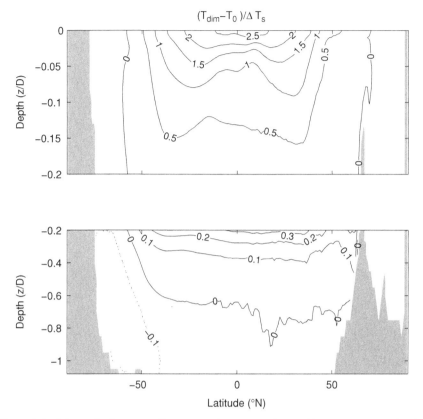

Figure 1.4. Dimensionless zonally averaged cross sections of long-term mean in situ temperature $T$ vs. latitude ($°$N) and depth $z/D$ for the world ocean, for dimensionless $T$ defined as in Figure 1.1 and $D = 5000$ m: (top) $-0.2 < z/D < 0$; (bottom) $z/D < -0.2$. The vertical coordinate $z$ is defined as positive upward, with $z = 0$ at the sea surface so that the ocean is confined to $z < 0$.

combination of two fundamentally different elements: an essentially deductive set of large-scale equations and a supplemental set of models or assumptions describing the mean effect of smaller-scale motion on the large scale. The approach taken here is to proceed deductively as far as possible and then to introduce the minimal set of supplemental models or assumptions required to obtain theories that are sufficiently complete to offer the essential insights into ocean circulation that are sought.

## 1.2 Physical Variables

In the continuum representation of the substance and flow of seawater that is the appropriate starting point for theories of large-scale ocean circulation, there are seven basic physical fields of interest, each of which is a function of time $t$ and of position $\mathbf{x}$ in the three-dimensional space in which the ocean resides. These physical fields are mass density $\rho$, three-component vector velocity $\mathbf{v}$, pressure $p$, thermodynamic

$(S_{pss} - S_0)/\Delta S$

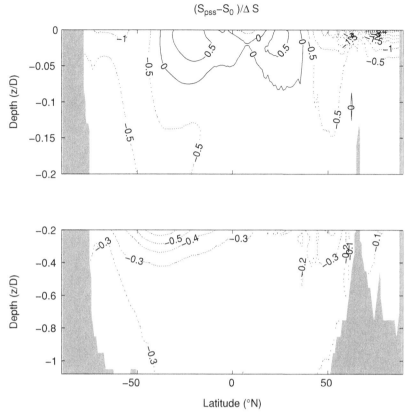

Figure 1.5. Dimensionless zonally averaged cross sections of long-term mean salinity $S$ vs. latitude (°N) and depth $z/D$ for the world ocean, for dimensionless $S$, defined as in Figure 1.2, and $z/D$, defined as in Figure 1.4: (top) $-0.2 < z/D < 0$; (bottom) $z/D < -0.2$.

energy $\hat{e}$, and salinity $S$. From a mathematical point of view, these physical fields form a complete set of dependent variables that describes the ocean state.

The fluid velocity **v** is itself best defined physically as a mass-weighted mean velocity of an infinitesimal fluid parcel. Two different pressures may be defined: the dynamic pressure and the thermodynamic pressure; following a standard approximation, it is assumed here that the fluid is locally in thermodynamic equilibrium and that these two pressures are equivalent. The thermodynamic energy $\hat{e}$ is a function of the pressure $p$, temperature $T$, and salinity $S$, or an equivalent set of three thermodynamic variables; this function is the equation of state for seawater and encodes the physical properties of the fluid substance. The salinity $S$ represents the concentration of a mixture of chemical salts whose relative proportions are set largely by chemical weathering of exposed continental forms and may be taken as uniform throughout the world ocean. In some cases, it can prove useful to exchange dependent and independent variables; for example, it can be convenient to use pressure as a pseudo-vertical coordinate.

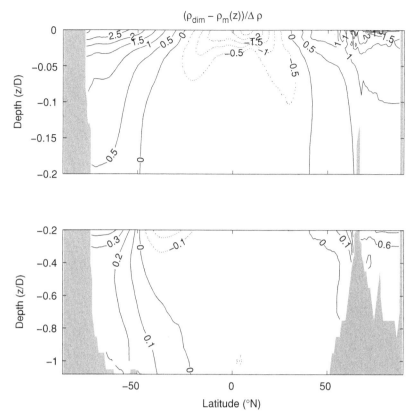

Figure 1.6. Dimensionless zonally averaged cross sections of long-term mean in situ density $\rho = [\rho_{\mathrm{dim}} - \rho_m(z)]/\Delta\rho$, defined as the departure from its horizontally averaged mean at each depth $[\rho_m(z)]$, scaled by $\Delta\rho = 1$ kg m$^{-3}$, vs. latitude ($^\circ$N) and depth $z/D$ for the world ocean, for $z/D$ as in Figure 1.4: (top) $-0.2 < z/D < 0$; (bottom) $z/D < -0.2$.

The equations that form the basis for the circulation theories are derived from conservation, or balance, principles for mass, momentum, salt, and thermodynamic energy, plus the empirically determined equation of state for seawater. These classical physical principles and the corresponding equations are briefly reviewed in the following sections. The corresponding derivations can be found in standard fluid mechanics texts.

## 1.3 Fluid Motion and the Material Derivative

The motion of a fluid in two- or three-dimensional space may be described by a two- or three-dimensional vector function or map $\mathbf{X}(\mathbf{a}, \tau)$ that gives, for each $\mathbf{a}$ in the domain, the position $\mathbf{x} = \mathbf{X}(\mathbf{a}, \tau)$ at time $\tau$ of the fluid parcel with initial position $\mathbf{x} = \mathbf{a}$ at $\tau = 0$. This is the so-called *Lagrangian* description, in which labeled fluid

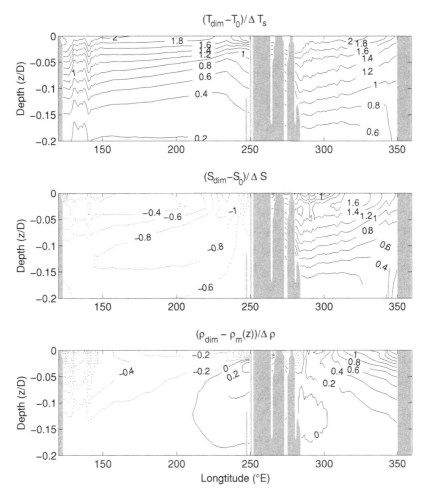

Figure 1.7. Dimensionless meridionally averaged cross sections of long-term mean in situ temperature, salinity, and density $\rho = [\rho_{\dim} - \rho_m(z)]/\Delta\rho$ vs. longitude (°E) and depth $z/D$ for the northern hemisphere subtropics (15°N–45°N) and $-0.2 < z/D < 0$. The dimensionless $T$, $S$, and $\rho$ are defined as in Figures 1.1, 1.2, and 1.6, respectively, and $z/D$ is defined as in Figure 1.4.

parcels are followed as they move. The Lagrangian fluid velocity $\mathbf{v}$ is then just the partial derivative of the map $\mathbf{X}$ with respect to time $\tau$, holding the initial positions $\mathbf{a}$ fixed:

$$\mathbf{v}(\mathbf{a}, \tau) = \frac{\partial}{\partial\tau}\mathbf{X}(\mathbf{a}, \tau). \qquad (1.1)$$

The inverse $\mathbf{A}(\mathbf{x}, t)$ of $\mathbf{X}$ gives the initial position $\mathbf{a}$ of the fluid parcel with position $\mathbf{x}$ at time $\tau = t$:

$$\mathbf{A}(\mathbf{x}, t) = \mathbf{A}[\mathbf{X}(\mathbf{a}, \tau = t), t] = \mathbf{a}. \qquad (1.2)$$

Alternatively, the fluid motion may be described by specifying the fluid velocity directly, at each point in space and time. This is the *Eulerian* description, in which the motion is considered relative to a fixed reference frame, and no parcel labels are required. The corresponding Eulerian fluid velocity field $\mathbf{u}(\mathbf{x}, t)$ is simply related to (1.1) and (1.2):

$$\mathbf{u}(\mathbf{x}, t) = \mathbf{v}[\mathbf{A}(\mathbf{x}, t), t] = \frac{\partial \mathbf{X}}{\partial \tau}[\mathbf{A}(\mathbf{x}, t), t]. \tag{1.3}$$

Note that the partial derivative $\partial \mathbf{X}/\partial \tau$ in (1.3) is still taken with respect to $\tau$ while holding $\mathbf{a}$ fixed, as in (1.1), and not with respect to $t$ while holding $\mathbf{x}$ fixed. Both $t$ and $\tau$ represent the same time variable, and their values are the same; the reason for the different symbols is this distinction between the meaning of the partial derivatives.

It turns out that the dynamical equations for fluid motion can often be posed in the Eulerian framework, with no, or limited, specification of the Lagrangian pathways of fluid parcels. In that case, the fluid velocity $\mathbf{u}(\mathbf{x}, t)$ is most conveniently defined as the ratio of the momentum to mass density of an infinitesimal fluid parcel or, equivalently, as a mass-weighted mean velocity of the fluid parcel. A Lagrangian representation (1.1) consistent with this definition can be obtained by modifying the initial Lagrangian label field $\mathbf{a}$ so that mass is uniformly distributed in $\mathbf{a}$ space.

Consider a scalar fluid property $\hat{F}(\mathbf{a}, \tau)$, which might be the concentration of a dissolved substance such as salinity. The rate of change of the value of this property for a given fluid parcel is the partial derivative of $\hat{F}$ with respect to $\tau$ while holding $\mathbf{a}$ fixed, and is denoted $D\hat{F}/Dt$:

$$\frac{D\hat{F}}{Dt} = \frac{\partial}{\partial \tau} \hat{F}(\mathbf{a}, \tau). \tag{1.4}$$

Suppose that $\hat{F}$ is given, instead, in terms of its distribution $F(\mathbf{x}, t) = \hat{F}[\mathbf{A}(\mathbf{x}, t), t]$ with respect to $\mathbf{x}$ at time $t$. Then the derivative following the motion, as a function of $\mathbf{x}$ and $t$, can be computed using the chain rule:

$$\frac{D\hat{F}}{Dt} = \frac{D}{Dt} F(\mathbf{x}, t) = \left( \frac{\partial}{\partial \tau} F[\mathbf{X}(\mathbf{a}, \tau), \tau] \right)_{\tau=t} = \mathbf{u} \cdot \nabla F + \frac{\partial F}{\partial t}. \tag{1.5}$$

Here $\partial F/\partial t$ denotes the local rate of change of $F$ with spatial coordinates $\mathbf{x}$ held fixed, and $\nabla$ is the three-dimensional gradient operator in $\mathbf{x}$ space so that $\nabla F = \operatorname{grad} F$ is the $\mathbf{x}$ gradient of $F$. The operator $D/Dt$ defined by (1.4) or (1.5) is the material derivative: the rate of change of a fluid property following the motion of the fluid.

## 1.4 Mass Conservation

The principle of conservation of mass demands that fluid mass be neither created nor destroyed and relates the mass density to the motion. With the continuum representation of the fluid, this principle can be expressed as a partial differential equation that balances the rate of change of the mass density $\rho$ and the divergence of the mass flux $\rho\mathbf{u}$:

$$\partial\rho/\partial t + \nabla \cdot \rho\mathbf{u} = \partial\rho/\partial t + \mathbf{u} \cdot \nabla\rho + \rho\nabla \cdot \mathbf{u} = \frac{D\rho}{Dt} + \rho\nabla \cdot \mathbf{u} = 0. \tag{1.6}$$

Here $\nabla \cdot \mathbf{F} = \operatorname{div} \mathbf{F}$ is the divergence of a vector field $\mathbf{F}$. Equation (1.6) is the Eulerian expression of the principle of mass conservation.

In the Lagrangian setting, the principle of mass conservation is

$$\rho J_a = \rho_0. \tag{1.7}$$

Here $\rho_0$ is the initial mass density, assumed constant in $\mathbf{a}$ space, and $J_a$ is the Jacobian determinant:

$$J_a = \det\left(\frac{\partial\mathbf{X}}{\partial\mathbf{a}}\right) = \det\left(\left\{\frac{\partial X_i}{\partial a_j}\right\}\right), \tag{1.8}$$

where $\partial X_i/\partial a_j$ is the matrix of partial derivatives of fluid-parcel positions $\mathbf{X}(\mathbf{a}, \tau)$ with respect to their initial positions $\mathbf{a}$. Equation (1.7) follows from the me transformation rule of integral calculus.

## 1.5 Momentum Balance

Newton's third law of motion, a basic principle of classical physics, relates the rate of change of momentum of a particle to the sum of applied forces. In a reference frame rotating with angular velocity $\boldsymbol{\Omega}$, the corresponding statement for a fluid may be written

$$\rho\left(\frac{D\mathbf{u}}{Dt} + 2\boldsymbol{\Omega} \times \mathbf{u}\right) = -\nabla p - \rho\nabla\Phi_g - \nabla \cdot \mathbf{d}, \tag{1.9}$$

where $\Phi_g$ is a potential for the sum of the gravitational and centrifugal forces and $\nabla \cdot \mathbf{d}$ represents the component-wise divergence $\partial d_{ij}/\partial x_j$ of the deviatoric stress tensor $\mathbf{d}$. In Cartesian coordinates $\mathbf{x} = (x, y, z) = (x_1, x_2, x_3)$,

$$(\mathbf{d})_{ij} = d_{ij} = -2\mu\left(e_{ij} - \frac{1}{3}\delta_{ij}\nabla \cdot \mathbf{u}\right), \tag{1.10}$$

where

$$e_{ij} = \frac{1}{2}\left(\frac{\partial u_i}{\partial x_j} + \frac{\partial u_j}{\partial x_i}\right). \tag{1.11}$$

In (1.10), $\mu$ is the dynamic viscosity, and $\delta_{ij}$ is the Kronecker $\delta$ function.

The term in (1.9) proportional to $\boldsymbol{\Omega}$ is the *Coriolis force*. For the rotating Earth, $|\boldsymbol{\Omega}| = \Omega = 7.29 \times 10^{-5}$ s$^{-1}$, and it will be sufficient here to take $\Phi_g = gz$, where $z$ is the local vertical coordinate and $g = 9.81$ m s$^{-2}$ is the effective gravitational acceleration. The dynamic viscosity $\mu = 1.3 \times 10^{-3}$ kg m$^{-1}$ s$^{-1}$ for pure water at $10°$C, and is within a few percent of this value for seawater.

## 1.6 Salt Conservation

Like the total fluid mass, the dissolved salts in seawater can be neither created nor destroyed, provided that biological fluxes and other changes of state can be neglected, as they will be here. This means that the concentration of dissolved salts, or salinity, $S$ of a fluid parcel can change only by diffusion:

$$\frac{DS}{Dt} = k_S \nabla^2 S. \tag{1.12}$$

Detailed consideration of thermodynamics shows that thermal and pressure gradients can also cause salt diffusion; however, these effects are small and are neglected here. A characteristic value for the saline diffusivity $k_S$, which depends on temperature, is $k_S = 1.5 \times 10^{-9}$ m$^2$ s$^{-1}$. For a 0.125M saline solution diffusing into pure water at temperatures near 300 K, an empirical expression is $k_S = k_1 \exp(-k_2/T)$, with $k_1 = 4.20 \times 10^{-6}$ m$^2$ s$^{-1}$ and $k_2 = 2368$ K. Oceanographic salinity values are now generally given in the dimensionless units of the practical salinity scale, which are nearly equivalent to parts per thousand by mass.

## 1.7 Thermodynamic Energy Balance

The first law of thermodynamics relates the change in internal energy of a thermodynamic system to the work done and the applied heating. For a compressible fluid, this principle can be expressed in terms of the rate of change following the motion of the internal energy $\hat{e}$ and density $\rho$ (or equivalently, specific volume $1/\rho$) of a fluid parcel:

$$\rho \left[ \frac{D\hat{e}}{Dt} + p \frac{D}{Dt}\left(\frac{1}{\rho}\right) \right] = k_T \nabla^2 T + \chi + \rho \, Q_e. \tag{1.13}$$

Here $k_T$ is the thermal conductivity, $\chi = 2\mu \left[ \Sigma_{ij} e_{ij} e_{ij} - (\nabla \cdot \mathbf{u})^2/3 \right]$ is internal mechanical heating from viscous dissipation, and $Q_e$ is external heating. For pure water at $10°$C, $k_T = 5.8 \times 10^{-1}$ J m$^{-1}$ s$^{-1}$ K$^{-1}$. A term representing the chemical potential for salinity could be included in (1.13) but is generally small and is neglected here.

## 1.8 Equation of State

The physical properties of seawater in thermodynamic equilibrium are described by an equation of state that relates one thermodynamic variable, usually the density $\rho$ or internal energy $\hat{e}$, to three others, usually pressure $p$, temperature $T$, and salinity $S$. In terms of density, the equation of state has the form

$$\rho = \mathcal{R}(p, T, S). \tag{1.14}$$

The physics of liquids remains mysterious, and a theoretical derivation of an equation of state from a molecular or quantum mechanical description of seawater is not currently available. Consequently, the function $\mathcal{R}$ must be determined empirically, which is generally done by fitting elementary functions in $p$, $T$, and $S$ to data from experimental measurements. The function $\mathcal{R}$ proves to be nonlinear, with the dominant nonlinearity being the dependence of thermal expansibility on pressure. Most values of in situ seawater density $\rho$ in the ocean satisfy $1020$ kg m$^{-3}$ $< \rho < 1070$ kg m$^{-3}$. A standard reference value is $\rho_0 = 1025$ kg m$^{-3}$. The specification of the equation of state in the form (1.14), rather than as an internal energy function, is convenient for oceanographic purposes. Measurement in the ocean of pressure, temperature, and salinity is much easier than direct measurement of density and, consequently, the equation of state (1.14) is used regularly to determine the dynamically important density from ocean observations.

## 1.9 Seawater Equations of Motion

In summary, the basic set of equations for a compressible, two-component, viscous, thermodynamic fluid subject to a gravitational force in a rotating coordinate system is

$$\frac{D\rho}{Dt} + \rho \nabla \cdot \mathbf{u} = 0, \tag{1.15}$$

$$\rho \left( \frac{D\mathbf{u}}{Dt} + 2\mathbf{\Omega} \times \mathbf{u} \right) = -\nabla p - \rho \nabla \Phi_g - \nabla \cdot \mathbf{d}, \tag{1.16}$$

$$\frac{DS}{Dt} = k_S \nabla^2 S, \tag{1.17}$$

$$\rho \left[ \frac{D\hat{e}}{Dt} + p \frac{D}{Dt} \left( \frac{1}{\rho} \right) \right] = k_T \nabla^2 T + \chi + \rho \, Q_e, \tag{1.18}$$

$$\rho = \mathcal{R}(p, T, S). \tag{1.19}$$

These represent, respectively, mass conservation, three components of momentum balance, salt conservation, thermodynamic energy balance, and the equation of state for seawater. This is a set of seven equations, which, when combined with suitable initial and boundary conditions, should, in principle, be sufficient to determine the seven unknown fields $\rho$, $\mathbf{u}$, $S$, $\hat{e}$, $p$; the equation of state, represented here by the empirically determined function $\mathcal{R}$ in (1.19), implicitly relates $\hat{e}$ to the temperature $T$ as the form of $\mathcal{R}$ indicates that any three thermodynamic variables are sufficient to determine the equilibrium thermodynamic state of a given fluid parcel. The parameters and fields $\boldsymbol{\Omega}$, $k_S$, $k_T$, $\Phi_g$, $\mathbf{d}$, $\chi$, and $Q_e$ are as defined in Sections 1.5–1.7, and the material derivative $D/Dt$ is as defined in (1.5).

## 1.10  Fluid Parcel Trajectories

Equations (1.15)–(1.19) are posed in the Eulerian framework, in which there is no explicit reference to Lagrangian labels $\mathbf{a}$ or trajectories $\mathbf{X}(\mathbf{a}, \tau)$. That it is possible to pose and solve the fluid equations in this simplified framework contrasts with the standard situation in classical mechanics, in which particle positions and momenta must, in general, be computed simultaneously. The existence of this possibility can be understood as a consequence of a parcel-exchange symmetry in the Hamiltonian formulation of fluid mechanics.

If, for a given set of initial and boundary conditions, a solution to (1.15)–(1.19) is found, the velocity field $\mathbf{u}(\mathbf{x}, t)$ is then known. The trajectories $\mathbf{X}(\mathbf{a}, \tau)$ can be obtained a posteriori by rewriting (1.3) as a set of ordinary differential equations for the maps $\mathbf{X}$,

$$
\frac{d\mathbf{X}}{d\tau} = \mathbf{u}[\mathbf{X}(\tau; \mathbf{a}), \tau], \tag{1.20}
$$

and solving these equations with the labels $\mathbf{a}$ as initial conditions. Because the labels $\mathbf{a}$ appear in this context only as initial conditions, it is consistent to write the time derivative in (1.20) as a total rather than partial derivative and to regard (1.20) as a set of ordinary rather than partial differential equations. In (1.20), the notation $\mathbf{X}(\tau; \mathbf{a})$ emphasizes that $\mathbf{X}$ has only parametric dependence on the initial conditions $\mathbf{a}$. The time variable is written as $\tau$ in (1.20) for explicit consistency with the distinction made in Section 1.3 between the Lagrangian and Eulerian independent variables $(\mathbf{a}, \tau)$ and $(\mathbf{x}, t)$: the trajectories are labeled by their initial conditions $\mathbf{a}$, and this label is held fixed for each trajectory as the time derivative is taken. The resulting instantaneous fluid parcel velocity $d\mathbf{X}/d\tau$ at $(\mathbf{x}, t = \tau)$ is equated in (1.20) to the local velocity vector of the fluid flow field $\mathbf{u}(\mathbf{x}, t = \tau)$.

Given a velocity field $\mathbf{u}(\mathbf{x}, t)$, it is always possible to compute solutions for the trajectories $\mathbf{X}(\tau; \mathbf{a})$ from (1.20). Even for relatively simple three-dimensional and time-independent, or two-dimensional and time-dependent, velocity fields, the resulting fluid trajectories can be extremely complicated. In addition, if the composition of the fluid parcel is subject to change by diffusion of material properties, such as salinity, according to (1.17), the physical meaning of these trajectories must be considered carefully, because the fluid parcel may eventually lose all practical identity as a discrete physical object. An equivalent ambiguity can arise when the velocity field $\mathbf{u}(\mathbf{x}, t)$ is taken to represent only the coherent, slowly varying, large-scale component of a complex flow with many scales of motion, and the average effects of the smaller-scale, rapidly fluctuating motions are presumed to act as a generalized diffusion on the large-scale flow.

## 1.11 Notes

A standard introduction to fluid dynamics, which gives a more detailed development of most of the material in this chapter, is Batchelor (1967); many of the values provided here for physical constants and parameters have been taken from that text. Additional material on the thermodynamics of seawater can be found in Fofonoff (1962) and Gill (1982). A new international standard equation of state has recently been adopted (IOC et al., 2010), in which salinity appears in units of $g \, kg^{-1}$, rather than on the practical salinity scale. The empirical expression for the dependence of $k_S$ on temperature in Section 1.6 is from Martinez et al. (2002). A discussion of the effect of temperature on salinity diffusion and an accessible development of Hamiltonian formulations of the fluid equations can both be found in the work of Salmon (1998). Historically, both the Eulerian and Lagrangian frameworks for the mathematical description of fluid motion should be attributed to L. Euler, with some credit due also to d'Alembert (Truesdell, 1954). Samelson and Wiggins (2006) discuss the complicated Lagrangian motion that can occur in geophysical flows with simple Eulerian velocity fields.

# 2

# Reduced Equations for Large-Scale Motion

## 2.1 Scaling

As a model of the fluid motion and thermodynamics of the Earth's oceans, the basic equations (1.15)–(1.19) include a formidable array of distinct types of physical processes, from divergence of molecular diffusive fluxes on scales of millimeters to coherent fluid flow on the scales of the Earth's circumference, a range of space scales of order $10^{10}$. This range of scales is comparable to that between the molecular scale and the scale of a mammal's body. Thus, in a rough sense, the problem of using these equations to understand the large-scale structure of the ocean is comparable in difficulty to using numerical computations of molecular interactions to simulate the behavior of a mammal. Numerical solution of this full set of equations is thus well beyond current computational capacities and will remain so for the foreseeable futuremore. Furthermore, these equations are sufficiently challenging that fundamental mathematical properties, such as the existence and uniqueness of solutions, are not established. Some of these basic properties remain a subject of mathematical research even for the incompressible Navier-Stokes equations, a simplified set of four equations that may be obtained from (1.15) and (1.16) by replacing the density variable $\rho$ with a constant value $\rho_0$.

Despite these formidable challenges, it is possible to make progress by using a combination of mathematical and physical reasoning to derive simplified, or reduced, equation sets that describe motions on the largest space and time scales of interest. Such progress is possible because these equations concisely express a robust set of physical principles that have validity over the entire vast range of scales, allowing the relative importance of individual terms in the equations to be estimated in a direct manner for motions of specific space and time scales. Through the use of such a priori estimates, a wide range of motions at smaller space and time scales can be effectively filtered out of the reduced equations for large-scale flow. The reduced equations are then amenable to direct analytical or numerical solution.

Simplified theoretical or empirical representations of the average effect of the small-scale motions on the large-scale flow can be subsequently inserted into the reduced, large-scale equations to obtain at least a qualitative understanding of the fundamental interactions between these scales of motion. Such closure theories are necessary, as one of the remarkable facts of the dynamics of ocean circulation is that the massive circulation systems that span the global ocean, and determine its thermal and haline structure on the very largest scales, are—even within the framework of the continuum fluid approximation—ultimately driven and mediated by physical processes that occur at the very smallest scales.

Direct observations of ocean physical variables suggest the following choices of characteristic scales for large-scale ocean circulation:

$$
\begin{aligned}
L &= \text{horizontal distance} &&= 5 \times 10^6 \text{ m} \\
D &= \text{vertical distance} &&= 5 \times 10^3 \text{ m} \\
T_0 &= \text{reference temperature} &&= 275 \text{ K} \\
S_0 &= \text{reference salinity} &&= 35 \\
\Delta T_s &= \text{surface temperature difference} &&= 10 \text{ K} \\
\Delta T_d &= \text{deep temperature difference} &&= 1 \text{ K} \\
\Delta S &= \text{salinity difference} &&= 1 \\
\rho_0 &= \text{reference density} &&= 1025 \text{ kg m}^{-3} \\
\Delta \rho_0 &= \text{in situ density difference} &&= 30 \text{ kg m}^{-3} \\
\Delta \rho &= \text{dynamic density difference} &&= 1 \text{ kg m}^{-3} \\
\Delta \zeta &= \text{sea-surface deformation} &&= 1 \text{ m} \\
W &= \text{vertical velocity} &&= 10^{-6} \text{ m s}^{-1} \\
U &= \text{horizontal velocity} &&= 10^{-3} \text{ m s}^{-1} \\
t_{\text{adv}} &= \text{time} &&= 5 \times 10^9 \text{ s}
\end{aligned}
\tag{2.1}
$$

The horizontal and vertical length scales $L$ and $D$ are taken from the dimensions of the ocean basins. The temperature and salinity scales may be obtained from ocean observations (Figure 2.1). The density scales are obtained from the temperature, salinity, and pressure observations and the empirical equation of state (1.19). The larger density scale, $\Delta \rho_0$, includes the mean compressibility effect of pressure on density, which dominates in situ density variations (Figure 2.1) but has no effect on horizontal pressure gradients (Section 2.3). The smaller density scale, $\Delta \rho$, characterizes the amplitude of the dynamically important lateral density gradients. The scale for sea-surface variations may be taken from satellite altimeter measurements or from the hydrostatic balance (Section 2.3). The scale $W$ for the vertical velocity, which is sufficiently small that it is difficult to measure directly, may be computed from Ekman boundary layer theory (Section 3.5) and observed distributions of midlatitude surface winds or wind stress. The horizontal velocity scale may be estimated as $U = WL/D$, anticipating a simplification of the mass conservation equation (1.15)

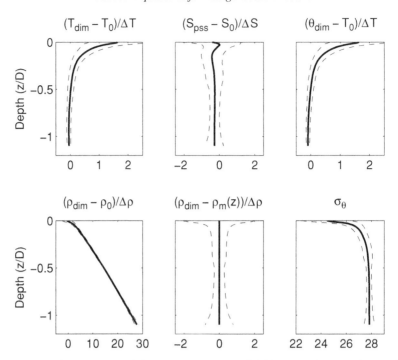

Figure 2.1. Volume-averaged long-term mean (solid lines) and mean $\pm$ standard deviation (dashed lines) profiles of dimensionless in situ temperature $T$, salinity $S$, potential temperature $\theta$, in situ density $\rho$ relative to the constant value $\rho_0$ and the horizontal mean at each depth $\rho_m(z)$, and potential density $\sigma_\theta$ vs. depth $(z/D)$ for the world ocean. The standard deviations are computed from differences from the horizontal mean at each level from a $1° \times 1°$ latitude-longitude gridded data set of long-term mean observed temperature and salinity. The dimensionless temperature, salinity, density deviations, and depth are defined as in Figures 1.4–1.7, using the same values of $T_0$ and $\Delta T$ for the dimensionless $T$ and $\theta$. The values of the dimensional scales $T_0$, $\Delta T$, $S_0$, $\Delta S$, $\rho_0$, and $\Delta \rho$ are from (2.1), with $\Delta T = \Delta T_s$ and $\Delta \rho = 1$ kg m$^{-3}$, and $\sigma_\theta = \rho_\theta - 1000$, with $\rho_\theta$ in units of kg m$^{-3}$.

to the incompressible form $\nabla \cdot \mathbf{u} = 0$. This horizontal velocity scale is an order of magnitude too small for the large-scale upper ocean circulation (Section 5.3) but is sufficient for the present purposes. The time scale $t_{\mathrm{adv}} = L/U$ is the advective time scale; its value, $5 \times 10^9$ s, is approximately equal to 160 years.

## 2.2 Spherical Polar Coordinates

The Earth, because of the balance of its gravitational self-attraction with the centrifugal force arising from its diurnal rotation, has the approximate shape of an oblate spheroid, with mean radius roughly 20 km greater at the equator than at the poles. To the accuracy of the theory considered in the following, it is adequate to idealize this shape as a sphere, with a gravitational and centrifugal force potential $\Phi_g$ in (1.16) that depends only on radial distance. Thus the natural coordinates are the spherical polar set $(\lambda, \phi, r)$, where $\lambda$ is longitude, $\phi$ is latitude, and $r$ is the radial and local vertical coordinate. The potential $\Phi_g$ is then taken to depend only on $r$. In an additional, traditional simplification, any deviation of the Earth's rotation vector $\mathbf{\Omega}$ from direction along the axis $\phi = \pi/2$ and constant magnitude is neglected.

In the spherical polar coordinate system, the components of the velocity $\mathbf{u}$ relative to the Earth are

$$\mathbf{u} = (u^\lambda, u^\phi, w). \tag{2.2}$$

To write the equations of motion (1.15)–(1.19) in spherical polar coordinates, the corresponding differential expressions must be obtained. The gradient of a scalar field $F(\lambda, \phi, r)$ is

$$\nabla F = \left( \frac{1}{r \cos\phi} \frac{\partial F}{\partial \lambda}, \frac{1}{r} \frac{\partial F}{\partial \phi}, \frac{\partial F}{\partial r} \right). \tag{2.3}$$

The divergence and curl of a vector field $\mathbf{F} = (F^\lambda, F^\phi, F^r)$ are, respectively,

$$\nabla \cdot \mathbf{F} = \frac{1}{r \cos\phi} \frac{\partial F^\lambda}{\partial \lambda} + \frac{1}{r \cos\phi} \frac{\partial (F^\phi \cos\phi)}{\partial \phi} + \frac{1}{r^2} \frac{\partial (r^2 F^r)}{\partial r} \tag{2.4}$$

$$\nabla \times \mathbf{F} = (\hat{F}^\lambda, \hat{F}^\phi, \hat{F}^r), \tag{2.5}$$

where

$$\hat{F}^\lambda = \frac{1}{r} \left[ \frac{\partial F^r}{\partial \phi} - \frac{\partial (r F^\phi)}{\partial r} \right], \tag{2.6}$$

$$\hat{F}^\phi = \frac{1}{r} \left[ \frac{\partial (r F^\lambda)}{\partial r} - \frac{1}{\cos\phi} \frac{\partial F^r}{\partial \lambda} \right], \tag{2.7}$$

$$\hat{F}^r = \frac{1}{r \cos\phi} \left[ \frac{\partial F^\phi}{\partial \lambda} - \frac{\partial (F^\lambda \cos\phi)}{\partial \phi} \right]. \tag{2.8}$$

The material derivative $DF/Dt$ of a scalar field $F(\lambda, \phi, r)$ is then, from (1.5),

$$\frac{DF}{Dt} = \frac{\partial F}{\partial t} + \frac{u^\lambda}{r \cos\phi} \frac{\partial F}{\partial \lambda} + \frac{u^\phi}{r} \frac{\partial F}{\partial \phi} + w \frac{\partial F}{\partial r}. \tag{2.9}$$

The three components of the acceleration $D\mathbf{u}/Dt$, the material derivative of the vector velocity, in the momentum equation (1.16) are

$$\frac{Du^\lambda}{Dt} = \frac{\partial u^\lambda}{\partial t} + \mathbf{u} \cdot \nabla u^\lambda + \frac{u^\lambda w}{r} - \frac{u^\lambda u^\phi}{r} \tan\phi, \tag{2.10}$$

$$\frac{Du^\phi}{Dt} = \frac{\partial u^\phi}{\partial t} + \mathbf{u} \cdot \nabla u^\phi + \frac{u^\phi w}{r} + \frac{(u^\lambda)^2}{r} \tan\phi, \tag{2.11}$$

$$\frac{Dw}{Dt} = \frac{\partial w}{\partial t} + \mathbf{u} \cdot \nabla w - \frac{(u^\lambda)^2 + (u^\phi)^2}{r}. \tag{2.12}$$

The corresponding components of the Coriolis force $2\mathbf{\Omega} \times \mathbf{u}$ in the momentum equation (1.16) are

$$2\mathbf{\Omega} \times \mathbf{u} = 2\Omega \left( w\cos\phi - u^\phi \sin\phi, \ u^\lambda \sin\phi, \ -u^\lambda \cos\phi \right). \tag{2.13}$$

It is useful to make the transformation

$$z = r - R_e + z_0, \tag{2.14}$$

where $R_e$ is the radius of the Earth and $z_0$ is chosen so that $z = 0$ at a convenient radius near mean sea level. A representative value of $R_e$ is the volumetric mean radius $R_e = 6371$ km. Because

$$\frac{\max\{|z|, |z_0|\}}{R_e} \approx \frac{D}{R_e} \approx 10^{-3} \ll 1, \tag{2.15}$$

an adequate approximation to this transformation may be made by replacing $r$ with $R_e$ and $\partial/\partial r$ with $\partial/\partial z$ in (2.3)–(2.12), before differentiating with respect to $r$. The resulting neglect of terms of the form $F/R_e$, relative to terms of the form $\partial F/\partial z$, in (2.4), (2.6), and (2.7) is consistent, provided that $F$ and the variation of $F$ over the ocean depth are of comparable magnitude. Thus, for example, the divergence (2.4) and material derivative (2.9) may be approximated respectively by,

$$\nabla \cdot \mathbf{F} = \frac{1}{R_e \cos\phi} \frac{\partial F^\lambda}{\partial \lambda} + \frac{1}{R_e \cos\phi} \frac{\partial (F^\phi \cos\phi)}{\partial \phi} + \frac{\partial F^r}{\partial z} \tag{2.16}$$

and

$$\frac{DF}{Dt} = \frac{\partial F}{\partial t} + \frac{u^\lambda}{R_e \cos\phi} \frac{\partial F}{\partial \lambda} + \frac{u^\phi}{R_e} \frac{\partial F}{\partial \phi} + w \frac{\partial F}{\partial z}. \tag{2.17}$$

## 2.3 The Hydrostatic Approximation

### *Large-Scale Vertical Momentum Balance*

The fundamental approximation for large-scale ocean motion is the hydrostatic approximation, in which the vertical momentum equation is treated as if the fluid were at rest, resulting in a quasi-static balance between the vertical component of the pressure gradient and the effective gravitational force. This, and the succeeding approximations, can be systematically motivated by estimating a priori the relative magnitudes of individual terms in (1.15)–(1.19), using the scales in (2.1).

To the accuracy required here, the potential $\Phi_g$ is given by

$$\Phi_g(z) = gz, \tag{2.18}$$

where $z$ is the local vertical coordinate defined in (2.14). For internal density variations of order $\Delta\rho$ in (2.1), the corresponding force per unit volume in the vertical momentum equation has the magnitude

$$|\Delta\rho\, g| \approx 10 \text{ N m}^{-3}. \tag{2.19}$$

In a similar way, the acceleration $\rho\, Dw/Dt$ may be estimated from (2.1) as

$$\left| \rho \frac{Dw}{Dt} \right| \approx \rho_0 \left\{ \frac{W}{t_{\text{adv}}}, \frac{UW}{L}, \frac{W^2}{D} \right\} \approx 2 \times 10^{-13} \text{ N m}^{-3}. \tag{2.20}$$

Thus the ratio of these two terms is

$$\left| \frac{\rho\, Dw/Dt}{\Delta\rho g} \right| \approx \frac{\rho_0}{\Delta\rho} \frac{U^2}{gD} \frac{D^2}{L^2} \approx 10^{-14} \ll 1. \tag{2.21}$$

This ratio is exceedingly small. Indeed, it is smaller than the relativistic corrections to the mass density in the gravitational force term, which (based on the Earth's orbital velocity) would be of order $U^2_{\text{Earth}}/c^2_{\text{light}} = 10^{-8}$. Thus it is well justified to neglect the term $Dw/Dt$ in (1.16). Note the appearance in (2.21) of the squares of the Froude number $\text{Fr} = U/(gD)^{1/2}$ and aspect ratio $\delta_a = D/L$.

A corresponding estimate for the vertical component of the Coriolis force is

$$|\rho_0 2\Omega u^\lambda \cos\phi| \approx 1.5 \times 10^{-4} \text{ N m}^{-3}. \tag{2.22}$$

Thus the ratio of this term to the gravitational force is

$$\left| \frac{2\Omega u^\lambda \cos\phi}{\Delta\rho\, g} \right| \approx \frac{\rho_0}{\Delta\rho} \frac{U\Omega}{g} \approx 10^{-5} \ll 1, \tag{2.23}$$

and the Coriolis term may likewise be neglected, though with a lesser degree of certainty than the acceleration $Dw/Dt$.

The vertical component of the stress divergence $\nabla \cdot \mathbf{d}$ in (1.16) may be estimated as

$$|\mu\nabla^2 w| \approx \frac{\mu W}{D^2} \approx 4 \times 10^{-18} \text{ N m}^{-3}, \tag{2.24}$$

and the resulting ratio of this term to the gravitational force is

$$\left| \frac{\mu\nabla^2 w}{\Delta\rho\, g} \right| \approx \frac{\rho_0}{\Delta\rho} \frac{\nu}{UL} \frac{U^2}{gD} \approx \frac{\rho_0}{\Delta\rho} \frac{\nu}{WD} \frac{D^2}{L^2} \frac{U^2}{gD} \approx 10^{-16} \ll 1. \tag{2.25}$$

This ratio, too, is exceedingly small, and the viscous term may also be neglected. Note the appearance in (2.25) of the inverses of the Reynolds numbers $\mathrm{Re}_h = UL/\nu$ and $\mathrm{Re}_v = WD/\nu$, based, respectively, on the horizontal and vertical velocity and length scales. Here $\nu = \mu/\rho = 1.3 \times 10^{-6}\,\mathrm{m^2\,s^{-1}}$ is the kinematic viscosity.

It follows from these considerations that to a high degree of accuracy, the vertical momentum equation for large-scale ocean circulation may be replaced by the condition of hydrostatic balance:

$$\frac{\partial p}{\partial z} = -\rho\frac{d\Phi_g}{dz} = -g\rho. \tag{2.26}$$

The appearance of several dimensionless parameters has been noted in the estimates leading to (2.26). For the scales, these parameters have the following values:

$$
\begin{array}{llll}
\delta_a & = D/L & = \text{aspect ratio} & = 1 \times 10^{-3} \\
\mathrm{Fr} & = U/(gD)^{1/2} & = \text{Froude number} & = 5 \times 10^{-6} \\
\mathrm{Re}_h & = UL/\nu & = \text{Reynolds number} & = 4 \times 10^{9} \\
\mathrm{Re}_v & = WD/\nu & = \text{Reynolds number} & = 4 \times 10^{3}
\end{array}
\tag{2.27}
$$

It is remarkable that despite the large Reynolds numbers, the vertical momentum equation is well approximated by a condition of static balance. This is one measure of the extent to which the basic physical character of the large-scale ocean circulation differs from that of the fluid flows of our everyday experience.

### *Pressure Effect on Density*

The increase in density with depth caused by compression of seawater under hydro-static pressure accounts for the largest changes in large-scale density over most of the ocean (Figure 2.1). A useful approximate solution of (2.26) may be computed by replacing the exact equation of state (1.19) by an approximate, linearized form:

$$\bar{\rho} = \bar{\mathcal{R}}(p) = \rho_0\left[1 + \Gamma_\rho(p - p_0)\right], \tag{2.28}$$

with constant compressibility $\Gamma_\rho = (1/\rho_0)\partial\rho/\partial p = (1/\rho_0)d\bar{\mathcal{R}}/dp$ and no dependence on $T$ or $S$. For seawater, a representative value of the compressibility is $\Gamma_\rho = 5 \times 10^{-10}\,\mathrm{Pa^{-1}}$. The corresponding solution of (2.28) with $\bar{p}(z=0) = p_0$ is then

$$\bar{p}(z) = p_0 + \Gamma_\rho^{-1}\left[\exp\left(-g\rho_0\Gamma_\rho z\right) - 1\right], \tag{2.29}$$

or

$$\bar{p}(z) \approx p_0 - g\rho_0 z, \tag{2.30}$$

where (2.30) follows because the ocean depth scale $D$ is much less than the compressibility scale depth $1/(g\rho_0\Gamma_\rho) \approx 2 \times 10^5$ m for seawater. The corresponding density profile is

$$\bar{\rho}(z) = \rho_0 \exp\left(-g\rho_0\Gamma_\rho z\right) \approx \rho_0(1 - g\rho_0\Gamma_\rho z), \tag{2.31}$$

with the approximate equality again following from $g\rho_0\Gamma_\rho D \ll 1$. A similar, more accurate approximate solution can be obtained by including pressure dependence of the compressibility $\Gamma_\rho$. Alternatively, a mean density profile $\bar{\rho}(z) = \rho_m(z)$ that includes both the compressibility effect and any systematic variations of $T$ and $S$ with depth may be defined directly from observations and the corresponding mean pressure profile obtained from the integral of (2.26):

$$\bar{p}(z) = p_0 - g \int_0^z \bar{\rho}(z')dz'. \tag{2.32}$$

In either case, because the hydrostatic equation (2.26) is linear, the deviations $p'$ and $\rho'$ of pressure and density, respectively, from the approximate functions $\bar{p}(z)$ and $\bar{\rho}(z)$ satisfy the same hydrostatic balance:

$$\frac{\partial p'}{\partial z} = -g\rho', \quad p'(\mathbf{x}, t) = p(\mathbf{x}, t) - \bar{p}(z), \quad \rho'(\mathbf{x}, t) = \rho(\mathbf{x}, t) - \bar{\rho}(z). \tag{2.33}$$

Also, because $\bar{p}$ depends only on $z$, it follows that $\nabla_h \bar{p} = 0$, where $\nabla_h$ is the horizontal gradient operator, and therefore that the horizontal pressure gradients of $p$ and $p'$ are equal:

$$\nabla_h p = \nabla_h p'. \tag{2.34}$$

Thus the variations in density due solely to pressure compression have no direct effect on the horizontal momentum balance, which is influenced only by $\rho'$ and $p'$.

Over most of the ocean, the corresponding range of variation of the dynamical density deviation $\rho'$ is of order $\Delta\rho = 1$ kg m$^{-3}$, an order of magnitude less than that of the density itself (Figure 2.1). This is the reason for the use of $\Delta\rho$ in place of the larger in situ density-difference scale $\Delta\rho_0$ in the scaling estimates in this section. An associated a priori scale estimate,

$$\Delta p' = g \, \Delta\rho \, \frac{D}{4} \approx 10^4 \text{ N m}^{-2}, \tag{2.35}$$

for the variations in pressure $p'$ associated with the density changes $\Delta\rho$ over the upper quarter of the water column, where the largest density variations are generally found (Figure 2.1), can be computed from (2.33). The corresponding equivalent dynamic difference $\Delta\zeta$ in sea level is

$$\Delta\zeta = \frac{\Delta p'}{g\rho_0} \approx 1 \text{ m}. \tag{2.36}$$

Note that the full pressure $p$ must always be considered in the thermodynamic equation (1.18) and the equation of state (1.19).

## 2.4 The Boussinesq Approximation

### *Incompressibility*

If the scales (2.1) are used to estimate the terms in the mass conservation equation (1.15), the ratio between the material derivative of density and any individual term in the product of the density and the velocity divergence proves to be proportional to the ratio $\delta_\rho$ of the maximum in situ density variation to the characteristic density. For example, for the vertical divergence term $\rho(\partial w/\partial z)$,

$$\left| \frac{D\rho/Dt}{\rho(\partial w/\partial z)} \right| \approx \frac{\Delta\rho_0}{\rho_0} \frac{\{1/t_{\mathrm{adv}}, U/L\}}{W/D} = \frac{\Delta\rho_0}{\rho_0} = \delta_\rho, \qquad (2.37)$$

and the same is true for each of the horizontal divergence terms because these are estimated by $\rho_0 U/L = \rho_0 W/D$. Direct observations show that over most of the ocean, a maximum value of the fractional density difference is $\delta_\rho \approx 3 \times 10^{-2} \ll 1$ (Figure 2.1). Thus Equation (1.15) can be approximated by the condition of incompressibility

$$\nabla \cdot \mathbf{u} = \nabla_h \cdot \mathbf{u}_h + \frac{\partial w}{\partial z} = 0, \qquad (2.38)$$

where $\mathbf{u}_h$ is the horizontal vector velocity.

Because the largest systematic, large-scale density variations over most of the ocean are those caused by compression in response to hydrostatic pressure variations (Figure 2.1), the incompressibility approximation (2.38) may also be motivated by considering the ratio $D/(g\rho_0\Gamma_\rho)^{-1}$ of the ocean depth scale $D$ to the density scale depth associated with the compressibility of seawater. The approximate hydrostatic solution (2.31) for the density shows that

$$\delta_\rho \approx g\rho_0\Gamma_\rho D \ll 1. \qquad (2.39)$$

However, the Boussinesq approximation (2.38) to the mass conservation equation does not depend on the hydrostatic approximation but instead only on the more general condition $\delta_\rho \ll 1$.

When the hydrostatic approximation is made, the evolution equation for $w$ is removed. The incompressibility condition (2.38) then appears as a diagnostic relation for $w$: from (2.38), $w$ can be determined from $\mathbf{u}_h$ at each $t$ simply by integrating (2.38) vertically, given a suitable boundary condition. It is important to recognize also that Equation (2.38) does not imply that $D\rho/Dt = 0$ as the equality (2.38) is only a statement that the divergence of the velocity field is negligible relative to the individual terms in the velocity divergence sum. The equation for the material rate of change of density must instead be obtained from the thermodynamic energy balance and the equation of state.

### Inertial Mass Density

The a priori estimate $\delta_\rho \ll 1$ implies also that in the inertial terms of the horizontal components of the momentum balance (1.16), the mass density $\rho$ may be replaced by the reference density $\rho_0$ because

$$\left| \frac{\rho}{\rho_0} \right| = \left| 1 + \frac{\rho - \rho_0}{\rho_0} \right| \approx 1 + \delta_\rho \approx 1. \tag{2.40}$$

With this approximation, the horizontal components of the momentum equations (1.16) become

$$\rho_0 \left[ \frac{D\mathbf{u}_h}{Dt} + 2(\mathbf{\Omega} \times \mathbf{u})_h \right] = -\nabla_h p - (\nabla \cdot \mathbf{d})_h. \tag{2.41}$$

Here $2(\mathbf{\Omega} \times \mathbf{u})_h$ is the horizontal component of the Coriolis force. The full density $\rho$, or the deviations of $\rho$ from $\rho_0$, is retained in the gravitational force term in the hydrostatic balance (2.26); replacing $\rho$ by $\rho_0$ in that term leads only to a simple, horizontally uniform, hydrostatic solution of the form (2.29), with no lateral pressure gradients and no motion.

This Boussinesq approximation to the inertia does not depend on the hydrostatic approximation. In the nonhydrostatic case, the same approximation $\rho \approx \rho_0$ may be made in the inertial terms of the full vertical momentum equation, while again retaining density variations in the gravitational force term. In general, the Boussinesq approximation to momentum equations consists of replacing the density by a constant in the inertial terms but not in the gravitational force term.

### Thermodynamic Energy

#### Entropy and Pressure Heating

Let $\eta(p, T, S)$ be the specific entropy of seawater. Then (1.18) may be written as

$$T\frac{D\eta}{Dt} = \left[ \frac{D\hat{e}}{Dt} + p\frac{D}{Dt}\left( \frac{1}{\rho} \right) \right] = \tilde{Q}_e, \tag{2.42}$$

where $\tilde{Q}_e$ represents the sum of the diabatic terms divided by the density $\rho$. The material derivative of $\eta$ may be expanded to yield

$$\frac{D\eta}{Dt} = \frac{\partial\eta}{\partial p}\bigg|_{T,S} \frac{Dp}{Dt} + \frac{\partial\eta}{\partial T}\bigg|_{p,S} \frac{DT}{Dt} + \frac{\partial\eta}{\partial S}\bigg|_{p,T} \frac{DS}{Dt}. \tag{2.43}$$

With the Maxwell relation

$$\frac{\partial\eta}{\partial p}\bigg|_{T,S} = -\frac{\partial}{\partial T}\left( \frac{1}{\rho} \right)\bigg|_{p,S} = \frac{1}{\rho^2}\frac{\partial\rho}{\partial T}\bigg|_{p,S} \tag{2.44}$$

and the definitions

$$\alpha_T = \rho \frac{\partial}{\partial T}\left(\frac{1}{\rho}\right)\Big|_{p,S} = -\frac{1}{\rho}\frac{\partial \rho}{\partial T}\Big|_{p,S} \tag{2.45}$$

$$C_p = T\frac{\partial \eta}{\partial T}\Big|_{p,S} \tag{2.46}$$

of the thermal expansion coefficient $\alpha_T$, and specific heat at constant pressure $C_p$, along with the previous neglect of the chemical potential term, the material rate of change of thermodynamic energy per unit mass in (2.42) can be rewritten as

$$T\frac{D\eta}{Dt} = C_p\frac{DT}{Dt} - \frac{\alpha_T T}{\rho}\frac{Dp}{Dt}. \tag{2.47}$$

Note that except for the neglect of the chemical potential, Equation (2.47) contains no approximations.

The first term on the right-hand side of (2.47) is the rate of change following the motion of the heat content per unit mass, while the second form describes the heating that results from pressure changes following the motion. The ratio of these two terms expresses the relative importance of the pressure effect on temperature and may be estimated from the scales (2.1) and (2.35):

$$\left|\frac{(\alpha_T T/\rho)\,Dp/Dt}{C_p\,DT/Dt}\right| \approx \frac{\alpha_T T_0 \rho_0 g D}{\rho_0 C_p \Delta T} = \frac{T_0}{\Delta T}\frac{g\alpha_T D}{C_p} \approx \frac{T_0}{\Delta T} \times 10^{-3}, \tag{2.48}$$

where the representative values $\alpha_T = 1 \times 10^{-4}$ K$^{-1}$ and $C_p = 4.2 \times 10^3$ J kg$^{-1}$ K$^{-1}$ have been used. Thus the magnitude of this ratio depends on the value of $\Delta T$, for which (2.1) offer two choices.

*Large Thermal Variations*

For theories that focus on the upper thermocline circulation, for which the estimate $\Delta T = \Delta T_s = 10$ K in (2.1) is appropriate, the ratio (2.48) is

$$\frac{T_0}{\Delta T_s} \times 10^{-3} \approx 3 \times 10^{-2}. \tag{2.49}$$

In this case, it is adequate to make the coarsest approximation to (2.47),

$$T\frac{D\eta}{Dt} = C_p\frac{DT}{Dt}, \tag{2.50}$$

resulting in the simplified thermodynamic energy equation

$$\frac{DT}{Dt} = \frac{\tilde{Q}_e}{C_p}. \tag{2.51}$$

In this version of the Boussinesq approximation, the pressure effect on temperature is neglected, and the temperature variable is the in situ temperature $T$, which is conserved following the motion in the absence of diabatic effects. The neglect of the pressure effect on temperature is an additional approximation, supplementing the approximations to the mass conservation and horizontal momentum equations (2.38) and (2.41).

## 2.5 Pressure-Corrected Temperatures and Densities

### *Potential Temperature*

For theories that treat the deep circulation, for which the estimate $\Delta T = \Delta T_d = 1$ K in (2.1) is appropriate, the ratio (2.48) is less than unity but not negligible, and the pressure effect on temperature cannot be disregarded. In this case, it would be natural to use the entropy $\eta$ as an independent thermodynamic variable, in place of the temperature $T$. Then the thermodynamic energy equation becomes

$$\frac{D\eta}{Dt} = \frac{\tilde{Q}_e}{T}, \tag{2.52}$$

and the equation of state (1.19) must be written in terms of $\eta$, according to

$$\rho = \hat{\mathcal{R}}(p, \eta, S) = \mathcal{R}[p, T(p, \eta, S), S], \tag{2.53}$$

and similarly for all other thermodynamic quantities.

Although it is possible to work directly with the entropy $\eta$ and (2.52), it is traditional to define and use instead a pressure-compensated pseudotemperature variable, the potential temperature, which has an heuristic similarity with and more immediate physical connection to the familiar in situ temperature. For a fluid parcel at pressure $p$ with temperature $T$ and salinity $S$, consider the adiabatic, salinity-conserving displacement of the parcel to a reference pressure $p_r$. Because the displacement is adiabatic, the entropy $\eta$ does not change, and by the thermodynamic identity (2.47), the imagined displacement must satisfy the differential relation

$$dT = \frac{\alpha_T T}{\rho C_p} dp \tag{2.54}$$

at every point along the displacement path. For the general seawater equation of state (1.19), $\alpha_T$, $\rho$, and $C_p$ are empirically known functions of $p$, $T$ and $S$. Because $\eta$ and $S$ are constant along the path, (2.54) may be integrated directly to obtain the potential temperature $\theta$, where

$$\theta(\eta, S; p_r) = T(p, \eta, S) \exp\left[\int_p^{p_r} \frac{\hat{\alpha}_T(q; \eta, S)}{\hat{\mathcal{R}}(q; \eta, S) \hat{C}_p(q; \eta, S)} dq\right]. \tag{2.55}$$

Here $\hat{\alpha}_T$ and $\hat{C}$ are the equivalents of $\alpha_T$ and $C_p$, defined following (2.53), and the order of the arguments indicates that $\eta$ and $S$ appear in the integrand only as parameters.

Direct calculation shows that

$$\frac{\partial \theta}{\partial p} = \left( \frac{\partial T}{\partial p} - \frac{\hat{\alpha}_T T}{\hat{\mathcal{R}} \hat{C}_p} \right) \frac{\theta}{T} = 0 \tag{2.56}$$

because the partial derivative $\partial T / \partial p$ may be identified with the differential in (2.54), in which $\eta$ and $S$ appear as constant parameters. Thus, by construction, $\theta$ is independent of $p$. For a given choice of reference pressure $p_r$, the potential temperature $\theta$ so constructed depends only on the entropy $\eta$ and salinity $S$; the reversible effect of pressure on temperature has been removed. Each distinct choice of reference pressure $p_r$ gives a distinct variety of potential temperature. The standard oceanographic potential temperature $\theta$ is defined with $p_r = p_a \approx 0$, where $p_a \approx 10^5$ Pa $\ll \rho_0 g D$ represents atmospheric surface pressure.

For fixed $p_r$, the material rate of change of potential temperature is

$$\frac{D\theta}{Dt} = \frac{\partial \theta}{\partial \eta} \frac{D\eta}{Dt} + \frac{\partial \theta}{\partial S} \frac{DS}{Dt} = \frac{\partial \theta}{\partial \eta} \frac{\tilde{Q}_e}{T} + \frac{\partial \theta}{\partial S} k_S \nabla^2 S. \tag{2.57}$$

Thus the potential temperature $\theta(\eta, S; p_r)$ is materially conserved in the absence of diabatic and molecular diffusive effects. Note that Equation (2.55) for the potential temperature is a definition that is exact to the accuracy of the empirical equation of state. Equation (2.57) may be rewritten as a thermodynamic energy equation:

$$T\frac{D\eta}{Dt} = C_p^\theta \left( \frac{D\theta}{Dt} - \frac{\partial \theta}{\partial S} \frac{DS}{Dt} \right) = \tilde{Q}_e, \tag{2.58}$$

where

$$C_p^\theta = T \frac{\partial \eta}{\partial \theta} \bigg|_{S,p} \tag{2.59}$$

is a specific heat for the potential temperature. To close the thermodynamic equations in terms of $\theta$, the equation of state (1.19) and other thermodynamic quantities must be rewritten as functions of $\theta$, now also with parametric dependence on the particular choice of reference pressure $p_r$, following

$$\rho = \tilde{\mathcal{R}}(p, \theta, S; p_r) = \mathcal{R}\{p, T[p, \eta(\theta, S; p_r), S], S\}, \tag{2.60}$$

where $\eta(\theta, S; p_r)$ is obtained by inverting (2.55) for $\eta$ at each $S$ and $p_r$.

For seawater, the equation of state $\mathcal{R}(p, T, S)$ is typically given as a high-order polynomial fit to measured data, and the potential temperature function $\theta(\eta, S; p_r)$ does not have a simple form. However, the density $\rho$ and specific heat $C_p$ depend only weakly on pressure. Although the dependence of the thermal expansion coefficient

$\alpha_T$ on pressure is not weak, a useful first approximation to the potential temperature can be obtained by setting $\rho$, $C_p$, and $\alpha_T$ to representative, constant values, which yields

$$\theta(\eta, S; p_r) \approx T \exp\left(\frac{p_r - p}{\rho_0 g D_\theta}\right) \approx T \exp\left(\frac{z - z_r}{D_\theta}\right), \tag{2.61}$$

where

$$D_\theta = \frac{C_p}{g\alpha_T} \approx 4 \times 10^6 \text{ m} \tag{2.62}$$

is a potential-temperature scale depth. In (2.61), the approximate solution (2.29) has been used to convert pressure dependence to depth dependence. This hydrostatic form could be arrived at more directly by using (2.29) in (2.47) or (2.54), but the general form (2.55) does not depend on the hydrostatic approximation.

Because $D/D_\theta \approx 10^{-3} \ll 1$, the potential temperature (2.61) may be further approximated:

$$\theta(\eta, S; p_r) \approx T\left(1 + \frac{p_r - p}{\rho_0 g D_\theta}\right) \approx T\left(1 + \frac{z - z_r}{D_\theta}\right). \tag{2.63}$$

In the ocean, the difference between the potential temperature (with $p_r = z_r = 0$) and the in situ temperature of a given water parcel is generally no more than 0.5 K (Figure 2.2); note that this is essentially the same estimate made in (2.48). Thus the pressure effect on temperature is much smaller, relative to general oceanic thermal variations, than is the pressure effect on density relative to general oceanic density variations.

Oceanic values of the thermal expansion coefficient $\alpha_T$ range from less than $0.25 \times 10^{-4}$ K$^{-1}$ to more than $3.5 \times 10^{-4}$ K$^{-1}$, with the larger values occurring for larger pressures and temperatures. A more accurate approximation than (2.61) may be obtained by letting $\alpha_T \approx \alpha_0 + \alpha_1 p$ so that

$$\theta(\eta, S; p_r) \approx T \exp\left\{\frac{g}{C_p}\left[\alpha_0(z - z_r) - \frac{1}{2}\alpha_1 g\rho_0\left(z^2 - z_r^2\right)\right]\right\}, \tag{2.64}$$

or

$$\theta(\eta, S; p_r) \approx T\left\{1 + \frac{g}{C_p}\left[\alpha_0(z - z_r) - \frac{1}{2}\alpha_1 g\rho_0\left(z^2 - z_r^2\right)\right]\right\}, \tag{2.65}$$

with the constants $\alpha_0$ and $\alpha_1$ chosen as, for example,

$$\alpha_0 \approx [0.52 + 0.12\,(T - 273.15)] \times 10^{-4} \text{ K}^{-1} \tag{2.66}$$

$$\alpha_1 \approx 0.23 \times 10^{-11} \text{ K}^{-1} \text{ Pa}^{-1} \tag{2.67}$$

for $T < 283$ K and $S \approx 35$ (Figure 2.2). This approximation includes the effect on $\theta$ of the increase of $\alpha_T$ with pressure or depth and with temperature.

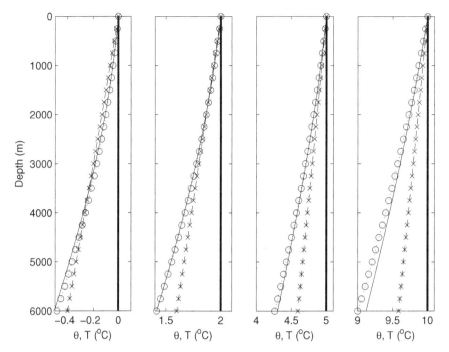

Figure 2.2. Profiles of exact (thin solid lines), approximate linear (crosses and dashed line; Equation (2.63)), and quadratic (circles and dotted line; Equation (2.64)) potential temperature $\theta$ (°C) vs. depth (m) for in situ $T = \{0, 2, 5, 10\}$°C (thick solid line).

### Gravitational Stability and Buoyancy Frequency

Consider a horizontally uniform fluid at rest, with vertical profiles $\theta = \Theta(z)$, $S = \mathcal{S}(z)$, $\rho = \bar{\rho}(z)$, and $p = \bar{p}(z)$ of potential temperature, salinity, density, and hydrostatic pressure, respectively. That is, $\bar{p}(z)$ is the solution of $d\bar{p}/dz = -g\bar{\rho}(z)$ with $\bar{p}(0) = p_a$, where $\bar{\rho}(z) = \tilde{\mathcal{R}}[\bar{p}(z), \Theta(z), \mathcal{S}(z)]$. Suppose that a fluid parcel is displaced vertically by a distance $\Delta z$ to the level $z_1$ and that the displacement is adiabatic and nondiffusive so that the parcel's potential temperature and salinity are unchanged. Let $\Delta\rho_1$ be the difference between the density of the displaced parcel and the density $\rho_1$ of the ambient fluid:

$$\Delta\rho_1 = \tilde{\mathcal{R}}[\bar{p}(z_1), \Theta(z_1 - \Delta z), \mathcal{S}(z_1 - \Delta z)] - \rho_1, \qquad (2.68)$$

where $\rho_1 = \bar{\rho}(z_1) = \tilde{\mathcal{R}}[\bar{p}(z_1), \Theta(z_1), \mathcal{S}(z_1)]$. Because the resulting perturbation to the pressure field will be small, of order $\Delta\rho_1/\rho_0$, the buoyant force on the displaced parcel at $z_1$ will be proportional to $\Delta\rho_1$. If $\Delta z$ is sufficiently small, the density difference may be approximated as

$$\Delta\rho_1 \approx \rho_1 \left( \alpha_\theta \frac{d\Theta}{dz} - \beta_S \frac{d\mathcal{S}}{dz} \right) \Delta z, \qquad (2.69)$$

where

$$\alpha_\theta = -\left.\frac{1}{\bar\rho}\frac{\partial\bar\rho}{\partial\theta}\right|_{p,S} \tag{2.70}$$

is the thermal expansion coefficient with respect to potential temperature and

$$\beta_S = \left.\frac{1}{\bar\rho}\frac{\partial\bar\rho}{\partial S}\right|_{p,\theta} \tag{2.71}$$

is a haline contraction coefficient. Thus, if $\alpha_\theta(d\Theta/dz) - \beta_S(dS/dz) > 0$, the density of an upward-displaced parcel will be greater than that of the ambient fluid, and the parcel will feel a downward buoyant force. In this case, the fluid is stably stratified because small perturbations to the profile are resisted. If $\alpha_\theta(d\Theta/dz) - \beta_S(dS/dz) < 0$, the density of an upward-displaced parcel will be less than that of the ambient fluid, and the parcel will feel an upward buoyant force. In this case, the fluid is convectively unstable because small perturbations to the profile are accelerated. Because $\theta \approx T$, the potential-temperature thermal expansion coefficient $\alpha_\theta \approx \alpha_T$, while oceanic values of $\beta_S$ are generally within 10 percent of $0.8 \times 10^{-3}$.

For the displaced parcel, the associated nonhydrostatic, linear vertical momentum balance is

$$\rho\frac{\partial w}{\partial t} \approx -g\Delta\rho_1. \tag{2.72}$$

With $\rho \approx \rho_1$ and $w \approx d(\Delta z)/dt$, this gives a differential equation for the parcel displacement of:

$$\frac{d^2}{dt^2}\Delta z = -N^2\Delta z, \tag{2.73}$$

where

$$N = \left[g\left(\alpha_\theta\frac{d\Theta}{dz} - \beta_S\frac{dS}{dz}\right)\right]^{1/2} \tag{2.74}$$

is the the buoyancy frequency. If $N^2 > 0$, Equation (2.73) gives oscillatory solutions for the displacement $\Delta z$, with frequency equal to $N$. The value of $N$, or of $N^2$, is thus a natural measure of the strength of the vertical stratification. If $N^2$ is to be computed from the vertical derivative of the density, rather than from the potential temperature and salinity distributions, the pressure effect on density must be removed:

$$N^2 = g\left(\alpha_\theta\frac{d\Theta}{dz} - \beta_S\frac{dS}{dz}\right) = -g\left(\frac{1}{\bar\rho}\frac{d\bar\rho}{dz} - \Gamma_\rho\frac{d\bar p}{dz}\right), \tag{2.75}$$

where the adiabatic compressibility coefficient $\Gamma_\rho$ is

$$\Gamma_\rho(p,\theta,S) = \left.\frac{1}{\bar\rho}\frac{\partial\bar\rho}{\partial p}\right|_{\theta,S} = \left.\frac{1}{\bar\rho}\frac{\partial\bar\rho}{\partial p}\right|_{\eta,S}. \tag{2.76}$$

Because $d\bar{p}/dz = -g\bar{\rho}$, the buoyancy frequency may also be written as

$$N^2 = -g\left(\frac{1}{\bar{\rho}}\frac{d\bar{\rho}}{dz} + \frac{g}{c_s^2}\right), \tag{2.77}$$

where the parameter $c_s = (\bar{\rho}\,\Gamma_\rho)^{-1/2}$ has dimensions of velocity (see Exercise 2.3).

A more rigorous treatment of these buoyancy oscillations requires consideration of the nonhydrostatic Boussinesq equations, linearized about a hydrostatic state of rest, with notation as in (2.33) and with the Coriolis terms neglected for motions of frequency much larger than $\Omega$:

$$\rho_0\frac{\partial \mathbf{u}_h}{\partial t} = -\nabla_h p', \quad \rho_0\frac{\partial w}{\partial t} = -\frac{\partial p'}{\partial z} - g\rho', \quad \nabla \cdot \mathbf{u} = 0, \tag{2.78}$$

with

$$\rho = \bar{\rho}(z) + \rho' = \tilde{\mathcal{R}}[\bar{p}(z) + p', \theta, S], \quad \frac{D\theta}{Dt} = 0, \quad \frac{DS}{Dt} = 0 \tag{2.79}$$

for adiabatic motions. Then the linearized, adiabatic material rate of change of density is

$$\frac{D\rho}{Dt} = \frac{\partial \rho'}{\partial t} + w\frac{d}{dz}\tilde{\mathcal{R}}[\bar{p}(z), \Theta(z), \mathcal{S}(z)] \tag{2.80}$$

or, also,

$$\frac{D\rho}{Dt} = \frac{\partial\tilde{\mathcal{R}}}{\partial p}\frac{Dp}{Dt} + \frac{\partial\tilde{\mathcal{R}}}{\partial\theta}\frac{D\theta}{Dt} + \frac{\partial\tilde{\mathcal{R}}}{\partial S}\frac{DS}{Dt} = \bar{\rho}\Gamma_\rho\frac{d\bar{p}}{dz}w, \tag{2.81}$$

where the last equality follows because $|\partial p'/\partial t|/|w(d\bar{p}/dz)| \approx (D/\bar{\rho})d\bar{p}/dz \approx \Delta\rho/\rho_0 \ll 1$. It follows that the local rate of change of density is given by

$$\frac{\partial\rho'}{\partial t} = \frac{\bar{\rho}}{g}N^2 w, \tag{2.82}$$

where $N^2$ is given by (2.75) and is assumed here to be constant. Eliminating $\mathbf{u}_h$, $p'$, and $\rho'$ in favor of $w$ then yields the linear internal wave equation:

$$\frac{\partial^2}{\partial t^2}\nabla^2 w + N^2\nabla_h^2 w = 0. \tag{2.83}$$

In Cartesian coordinates, this equation has solutions of the form

$$w = w_0\exp[-i(kx + ly + mz - \omega_{IW}t)], \tag{2.84}$$

with dispersion relation

$$\omega_{IW}^2 = \frac{k^2 + l^2}{k^2 + l^2 + m^2}N^2. \tag{2.85}$$

The maximum frequency of these oscillations, approached for $k^2 + l^2 \gg m^2$, is the buoyancy frequency $N$.

### *Potential Density*

By analogy with the potential temperature, a potential density, $\rho_\theta$, may be defined as the density that a fluid parcel would have if it were moved to a reference pressure $p_r$ adiabatically, with no change in salinity. Equivalently, the potential density is the density computed from the equation of state at the reference pressure, using the potential temperature that is defined with respect to the same reference pressure:

$$\rho_\theta(\eta, S; p_r) = \tilde{\mathcal{R}}[p_r, \theta(\eta, S; p_r), S]. \tag{2.86}$$

Like potential temperature, the potential density is materially conserved in adiabatic, nondiffusive flow because

$$\frac{D\rho_\theta}{Dt} = \frac{\partial \tilde{\mathcal{R}}}{\partial \theta} \frac{D\theta}{Dt} + \frac{\partial \tilde{\mathcal{R}}}{\partial S} \frac{DS}{Dt} \tag{2.87}$$

for a fixed choice of reference pressure $p_r$.

For given vertical profiles $\theta = \Theta(z)$ and $S = \mathcal{S}(z)$ of potential temperature and salinity, the vertical derivative of potential density is

$$\frac{\partial \rho_\theta}{\partial z} = -\rho_\theta \left( \alpha_\theta|_{p_r} \frac{\partial \Theta}{\partial z} - \beta_S|_{p_r} \frac{\partial \mathcal{S}}{\partial z} \right). \tag{2.88}$$

Because the expansion coefficient $\alpha_\theta$ and contraction coefficient $\beta_S$ in (2.88) are evaluated at the reference pressure $p_r$ rather than at the local pressure or vertical level, as in (2.74), the expressions in parentheses in (2.88) and (2.74) are equal only at the level of the reference pressure $p_r$. Thus the vertical gradient of potential density can only be used to compute the buoyancy frequency at the reference pressure $p_r$. Far from $p_r$, the vertical gradient of potential density can even differ from $N^2$ in sign and does not determine local stability. For this and other related reasons, the potential density is of limited theoretical use. A standard set of distinct potential density varieties with different reference pressures are frequently used in the analysis and presentation of ocean data so that a reference pressure near the pressures of interest for a particular analysis can be chosen. These potential densities are typically reported as dimensionless quantities $\sigma_\theta = \rho_\theta - 1000$, for values of $\rho_\theta$ in units of kg m$^{-3}$.

## 2.6 Inviscid, Nondiffusive, Adiabatic Flow

It can be anticipated from the large Reynolds numbers in (2.27) that for large-scale ocean circulation, the viscous terms associated with molecular diffusion of momentum may be neglected in the horizontal as well as vertical momentum balance. Although the inertial terms turn out not to be the largest terms in these balances, it is adequate for this purpose to compare the viscous terms to the inertial terms in the horizontal components of (1.16). This ratio yields the Reynolds number Re$_v$ from (2.27):

$$\left| \frac{\mu \nabla^2 \mathbf{u}_h}{\rho \mathbf{u} \cdot \nabla \mathbf{u}_h} \right| \approx \frac{\mu U \max \left\{ 1/D^2, 1/L^2 \right\}}{\rho_0 U \left\{ U/L, W/D \right\}} \approx \frac{\nu}{WD} = \frac{1}{\text{Re}_v} = 2 \times 10^{-4} \ll 1. \quad (2.89)$$

Thus the viscous terms $\nabla \cdot \mathbf{d}$ in the horizontal components of (1.16) may also be neglected.

The terms describing molecular diffusion of heat and salt in the thermodynamic energy and salinity equations may be estimated in a similar manner. For salinity (1.17),

$$\left| \frac{k_S \nabla^2 S}{\mathbf{u} \cdot \nabla S} \right| \approx \frac{k_S \Delta S/D^2}{W \Delta S/D} \approx \frac{k_S}{WD} = \frac{1}{\text{Pe}_S} = 3 \times 10^{-7} \ll 1, \quad (2.90)$$

where Pe$_S$ is a haline Peclét number. For thermodynamic energy (1.13),

$$\left| \frac{k_T \nabla^2 T}{\rho \mathbf{u} \cdot \nabla \hat{e}} \right| \approx \frac{k_T \Delta T/D^2}{\rho_0 C_p W \Delta T/D} \approx \frac{\kappa_T}{WD} = \frac{1}{\text{Pe}_T} = 3 \times 10^{-5} \ll 1, \quad (2.91)$$

where $\kappa_T = k_T/(\rho C_p) = 1.4 \times 10^{-7}$ m$^2$ s$^{-1}$ is the thermal diffusivity and Pe$_T$ is a thermal Peclét number. For the thermodynamic equation, the relative magnitude of the dissipative heating $\chi$ may also be estimated, giving

$$\left| \frac{\chi}{\rho \mathbf{u} \cdot \nabla \hat{e}} \right| \approx \frac{10 \nu U^2/D^2}{C_p W \Delta T/D} \approx \frac{10 \nu U L}{C_p D^2 \Delta T} = 5 \times 10^{-16} \ll 1, \quad (2.92)$$

where the smaller value $\Delta T = \Delta T_d$ from the scales given in Section 2.1 has been used.

Thus, for large-scale ocean motions with the scales given by in Section 2.1, the viscous, diffusive, and diabatic terms may be neglected; these are $(\nabla \cdot \mathbf{d})_h$ in (2.41), $k_S \nabla^2 S$ in (1.17) and (2.57), and $\tilde{Q}_e$ in (2.51) and (2.57). The additional approximation $Q_e = 0$ in (1.18) is thereby also made, which is appropriate away from a surface layer of order 50 m depth, over which solar radiative heating is distributed. The resulting absence of any sources of heating on the right-hand sides of (2.51) and (2.57) indicates that the flow is adiabatic: heat content per unit mass is conserved following the motion.

In the momentum, salinity, and thermodynamic equations, the neglect of the molecular diffusion terms means a reduction in the spatial derivative order of the equations. This, in turn, means a reduction in the number of spatial boundary conditions that can be imposed on the equations. In standard inviscid flow theory, for example, the condition of no flow at a rigid boundary must generally be relaxed to the less restrictive condition of no normal flow at the boundary, for which the tangential flow is unconstrained. The important issue of boundary conditions for reduced equations will reappear in several subsequent contexts.

## 2.7 The Primitive Equations for the Ocean

In the horizontal momentum equations (2.41), the ratio of the two Coriolis terms in the zonal momentum equation is

$$\left| \frac{\rho_0 \, 2\Omega \, w \, \cos \phi}{\rho_0 \, 2\Omega \, u^\phi \, \sin \phi} \right| \approx \delta_a \frac{\cos \phi}{\sin \phi}. \tag{2.93}$$

Because the aspect ratio $\delta_a \ll 1$, the term in $w$ may be neglected away from the equator, where $\phi = 0$. Near the equator, this term may be compared to the pressure term using (2.35) and the horizontal scale $L$ to estimate the horizontal pressure gradient:

$$\left| \frac{\rho_0 \, 2\Omega \, w \, \cos \phi}{\nabla_h p} \right| \approx \left| \frac{\rho_0 2\Omega W}{\Delta \rho g D/L} \right| \approx 1 \times 10^{-5}. \tag{2.94}$$

This motivates the neglect of the Coriolis term in $w$ also at the equator. Note, however, that an estimate, using the scales (2.1) and (2.35), of the ratio of the remaining term $\rho_0(\partial u^\lambda / \partial t)$ at the equator to the pressure gradient would similarly be small. This is a warning that the equatorial region has a special character and that its representation in large-scale circulation theory may require special attention, including the use of boundary layer methods.

With the neglect of the Coriolis term in $w$ and the viscous terms, the Boussinesq horizontal momentum equations (2.41) become

$$\frac{D\mathbf{u}_h}{Dt} + 2\Omega \sin \phi \, \mathbf{k} \times \mathbf{u}_h = -\frac{1}{\rho_0} \nabla_h p, \tag{2.95}$$

where $\mathbf{k}$ is the local vertical unit vector. Here, and elsewhere below, the notation $\mathbf{k} \times \mathbf{v}_h$ with a two-dimensional vector $\mathbf{v}_h = (v_1, v_2)$ should be interpreted as $\mathbf{k} \times \mathbf{v}_h = (-v_2, v_1)$. Equations (2.95), together with the hydrostatic equation (2.26),

$$\frac{\partial p}{\partial z} = -g\rho; \tag{2.96}$$

the Boussinesq mass conservation equation (2.38),

$$\nabla_h \cdot \mathbf{u}_h + \frac{\partial w}{\partial z} = 0; \tag{2.97}$$

the salinity conservation equation (1.17) with diffusion neglected,

$$\frac{DS}{Dt} = 0; \tag{2.98}$$

and one of the two pairs of adiabatic thermodynamic energy equations and equations
of state, either (2.51) and (1.19),

$$\frac{DT}{Dt} = 0, \quad \rho = \mathcal{R}(p, T, S), \tag{2.99}$$

or (2.57) and (2.60),

$$\frac{D\theta}{Dt} = 0, \quad \rho = \tilde{\mathcal{R}}(p, \theta, S; p_r), \tag{2.100}$$

form the hydrostatic, Boussinesq primitive equations for ocean circulation.

## 2.8 Thermohaline Planetary Geostrophic Equations

For the broad, slow, large-scale ocean flows characterized by the scales of Section 2.1,
the inertial term in the horizontal momentum equations is dominated by the Coriolis
force:

$$\left| \frac{D\mathbf{u}_h/Dt}{2\Omega \, \mathbf{u}_h \, \sin\phi} \right| \approx \frac{U\{1/t_{\text{adv}}, U/L\}}{2\Omega \, U \, \sin\phi} = \frac{U}{2\Omega \, L \, \sin\phi} = \epsilon \approx 1 \times 10^{-6}, \tag{2.101}$$

where it is assumed that $\sin\phi$ is of order unity. The dimensionless ratio $\epsilon =
U/(2\Omega L \sin\phi)$ is the Rossby number. Thus, away from the equator, the Rossby
number for large-scale ocean flows is very small, and the material derivative of the
velocity may be neglected in the horizontal momentum equations.

This results in a general form of the thermohaline planetary geostrophic equations
for large-scale ocean circulation:

$$f \, \mathbf{k} \times \mathbf{u}_h = -\frac{1}{\rho_0} \nabla_h p, \tag{2.102}$$

$$\frac{\partial p}{\partial z} = -g\rho, \tag{2.103}$$

$$\nabla_h \cdot \mathbf{u}_h + \frac{\partial w}{\partial z} = 0, \tag{2.104}$$

$$\frac{DS}{Dt} = 0, \tag{2.105}$$

and either

$$\frac{DT}{Dt} = 0, \quad \rho = \mathcal{R}(p, T, S) \tag{2.106}$$

or

$$\frac{D\theta}{Dt} = 0, \quad \rho = \tilde{\mathcal{R}}(p, \theta, S; p_r). \tag{2.107}$$

In (2.102), the Coriolis parameter, or planetary vorticity, $f$ has been defined:

$$f = 2\Omega \sin\phi. \tag{2.108}$$

The parameter $f$ is a function of latitude $\phi$ due to the meridional variation of the local vertical component of the Earth's rotation vector. The equality of the Coriolis and pressure gradient terms in the horizontal momentum equation (2.102) is the geostrophic balance. The geostrophic and hydrostatic balance conditions (2.102) and (2.103) are the basic dynamical approximations for large-scale motion.

These planetary geostrophic equations are thermohaline in that they include independent evolution equations for temperature and salinity and a full equation of state. In the planetary geostrophic dynamics, the temporal evolution of the ocean state is controlled by the advection of temperature, or potential temperature, and salinity by the three-dimensional velocity field. The resulting mass and pressure fields are related diagnostically to the temperature and salinity fields by the equation of state and the hydrostatic relation plus a boundary condition on the pressure. The horizontal velocity is related diagnostically to the pressure gradient through the geostrophic balance, and the vertical velocity is related diagnostically to the horizontal velocity through the mass conservation equation plus a boundary condition on the vertical velocity.

If a vertical derivative of the geostrophic balance equations is taken, and the hydrostatic balance is substituted, the result is the thermal wind balance:

$$f\,\mathbf{k} \times \frac{\partial \mathbf{u}_h}{\partial z} = \frac{g}{\rho_0}\nabla_h\rho \quad \text{or} \quad \frac{\partial \mathbf{u}_h}{\partial z} = -\frac{g}{\rho_0 f}\mathbf{k} \times \nabla_h\rho. \tag{2.109}$$

Thus, for a fluid in geostrophic balance, there will be geostrophic motion whenever there are lateral density gradients. If the geostrophic velocity is known at some level $z_1$, the thermal wind equations may be integrated vertically to obtain the geostrophic velocity at any depth $z$, from knowledge only of the mass field:

$$\mathbf{u}_h(z) = \mathbf{u}_h(z_1) - \frac{g}{\rho_0 f}\int_{z_1}^{z} \mathbf{k} \times \nabla_h\rho\,dz'. \tag{2.110}$$

By (2.33), vertically varying but horizontally uniform pressure or density fields have no effect on the geostrophic or thermal wind balances, and only the deviations from such horizontally uniform fields need be considered when computing the geostrophic velocity. Note here that as in the case of the buoyancy frequency, the potential density, again, cannot be used to compute the needed density gradients in (2.109), or (2.110), except at the potential-density reference level itself, at which the potential density is exactly equal to the in situ density.

The thermohaline planetary geostrophic equations (2.102)–(2.107) are a deductive set of equations that must be approximately satisfied by large-scale ocean flows with characteristic scales (2.1). As such, they provide a relatively reliable and well-motivated starting point for theories of the large-scale ocean circulation.

Because the characteristic scales were specified a priori, however, the resulting solutions may in principle violate the original scaling, and the consistency of the approximations may only be formally demonstrated a posteriori. Many differentiated terms were ultimately neglected in the derivation, and thus these simplified equations are also degenerate in the sense that regular solutions may be expected to exist at most for only certain special sets of boundary conditions. It must also be emphasized that the derivation has followed a filtering rather than an averaging approach: no account has been taken of the possibility that nonlinearities in the original equations may lead to integrated effects on the large-scale flow from correlated, fluctuating, smaller-scale or higher-frequency motions. Instead, only the balances explicitly appropriate for the intrinsic dynamics of the large-scale flow, as characterized by the specified scales in Section 2.1, are represented. With these numerous complications, it is clear that meaningful progress requires careful attention to the physical context in which solutions of these simplified equations are sought.

## 2.9 Notes

Spherical polar coordinate expressions for the rate-of-strain tensor $\{e_{ij}\}$ in (1.11), along with derivations of (2.3)–(2.12), are given by Batchelor (1967). Gill (1982) provides a useful discussion of thermodynamic equations and approximations. Basic forms of the planetary geostrophic equations were first derived independently by Robinson and Stommel (1959) and Welander (1959), with a similar scaling for atmospheric motions having been explored, also independently, by Burger (1958). The terminology *planetary geostrophic equations* was introduced by de Verdiére (1988). Systematic developments of these equations as large-scale approximations are given also in the texts by Pedlosky (1987) and Vallis (2006). General mathematical properties, such as existence, uniqueness, and regularity, of solutions of models based on the planetary geostrophic equations have been analyzed, for example, by Samelson et al. (2000) and Cao et al. (2004). Pedlosky (1996) gives an example of an equatorial boundary layer theory for the large-scale circulation.

# 3

# Planetary Geostrophic Vorticity Dynamics

## 3.1 Vorticity Balance

A planetary geostrophic equation for the vertical component of vorticity may be obtained by cross-differentiating the horizontal momentum equations (2.102), following (2.8):

$$
0 = \frac{1}{R_e \cos \phi} \left[ \frac{\partial}{\partial \lambda} \left( 2\Omega \, u^\lambda \, \sin \phi + \frac{1}{\rho_0 R_e} \frac{\partial p}{\partial \phi} \right) \right.
$$
$$
\left. - \frac{\partial}{\partial \phi} \left( -2\Omega \, u^\phi \, \sin \phi \, \cos \phi + \frac{1}{\rho_0 R_e} \frac{\partial p}{\partial \lambda} \right) \right]
$$
$$
= \frac{2\Omega \, \sin \phi}{R_e \cos \phi} \left( \frac{\partial u^\lambda}{\partial \lambda} + \frac{\partial (u^\phi \cos \phi)}{\partial \phi} \right) + \frac{2\Omega \, \cos \phi}{R_e} u^\phi. \tag{3.1}
$$

Using (2.104) and (2.108), (3.1) may be written as

$$
\beta \, u^\phi = f \frac{\partial w}{\partial z}, \tag{3.2}
$$

where

$$
\beta = \frac{1}{R_e} \frac{df}{d\phi} = \frac{2\Omega \, \cos \phi}{R_e} \tag{3.3}
$$

is the meridional gradient of the Coriolis parameter $f$. Note that here, $\beta$ is a function of latitude $\phi$.

Equation (3.2) is the Sverdrup vorticity relation. It is a diagnostic equation that relates the meridional velocity $u^\phi$ to the vertical divergence $\partial w / \partial z$ at each point in the domain and at each instant in time. Its physical meaning is that the change in vorticity arising from the stretching of vertical vorticity lines $f$ by the vertical divergence $\partial w / \partial z$ and the change in vorticity induced by meridional motion $u^\phi$ in the gradient $\beta$ of the vertical vorticity field $f$ must balance exactly at each point. This quasi-steady balance of two terms is an extremely restrictive form of the general fluid

37

vorticity equation and places strong constraints on the planetary geostrophic motion field.

From a deductive point of view, the vorticity balance (3.2) follows from the basic assumption that the large-scale flow is characterized by the a priori scale estimates (2.1); that is, if such a flow exists, then it must satisfy (3.2). This result can also be established directly from a more general vorticity equation, such as that derived from cross-differentiation of the primitive-equation horizontal momentum equations (2.95), by using the scales (2.1) to estimate the relative magnitudes of individual terms, as in the derivation of the planetary geostrophic equations (2.102)–(2.107) from the general fluid equations (1.15)–(1.19).

A useful heuristic interpretation of (3.2) can be constructed by recognizing that the general vorticity equation is essentially a statement of conservation of angular momentum and by idealizing the large-scale ocean circulation characterized by (2.1) as a laterally broad and vertically thin disk of fixed volume. The disk, with characteristic diameter $L$ and height $D$, is nearly at rest with respect to the Earth and so has angular momentum proportional to the local vertical component of the Earth's angular velocity. As the disk is stretched vertically by a vertical divergence $\partial w/\partial z$, the radius of the disk must change proportionally to conserve volume. The corresponding change in the moment of inertia might be expected to result, as a consequence of angular momentum conservation by the disk, in an altered rate of rotation of the disk relative to the Earth. However, because of the large horizontal scale $L$ of the motion, the induced azimuthal velocity at the outer portions of the disk would quickly become much larger than the a priori velocity scale $U$. Such a response is thus inconsistent with the given scale estimates (2.1). If, instead, the disk moves meridionally in such a way that its absolute angular velocity remains equal to the magnitude of the local vertical component of the Earth's rotation vector, it can conserve angular momentum without developing velocities relative to the Earth that are larger than $U$. According to the deductive equation (3.2), this balance of meridional motion and vertical divergence, calibrated through the ratio $f/\beta$, turns out to be the only vorticity balance consistent with the characteristic scales of the large-scale ocean circulation.

## 3.2 The $\beta$-Plane

Near a fixed latitude $\phi_0$, the horizontal components $(\lambda, \phi)$ of the spherical coordinate system may be replaced by Cartesian coordinates $(x, y)$, where

$$x = R_e \cos \phi_0 \, \lambda, \quad y = R_e(\phi - \phi_0). \tag{3.4}$$

The Coriolis parameter $f$ may also be represented locally in Cartesian coordinates as

$$f = f_0 + \beta \, (y - y_0), \tag{3.5}$$

where $f_0 = 2\Omega \sin \phi_0$ and

$$\beta = \frac{2\Omega \cos \phi_0}{R_e} \qquad (3.6)$$

is now a constant. These local Cartesian coordinates are known as the β-plane because they represent a tangent-plane approximation to the spherical surface at latitude $\phi_0$, within which the meridional gradient of the Coriolis parameter $f$ is retained and represented by the constant $\beta$. In these coordinates, the horizontal gradient and divergence operators in (2.102) and (2.104) are based on the standard Cartesian form $\nabla_h = (\partial/\partial x, \partial/\partial y)$. Characteristic midlatitude values of the β-plane constants $f_0$ and $\beta$ in (3.5) are $f_0 = 10^{-4}$ s$^{-1}$ and $\beta = 2 \times 10^{-11}$ m$^{-1}$ s$^{-1}$.

In quasi-geostrophic scaling, which leads to the quasi-geostrophic rather than planetary geostrophic equations, the meridional scale $L_{QG}$ of the motion is limited such that $L_{QG} \ll f_0/\beta$. In that case, the β-plane coordinates can be motivated as an asymptotic, local approximation to the spherical coordinates. The β-plane provides several convenient simplifications, including the absence of metric terms in the differential operators and the representation of the parameter $\beta$ as a constant. These simplifications make the β-plane particularly appealing as a framework for theoretical analysis of ocean circulation.

In the planetary geostrophic case of interest here, $L \approx f_0/\beta$, and, unfortunately, the use of the β-plane approximation cannot be motivated as an asymptotic approximation. However, it turns out that the dynamics of the planetary geostrophic equations are not essentially different on the β-plane and in spherical coordinates. In particular, the fundamental planetary geostrophic vorticity equation (3.2) has an exactly analogous form on the β-plane: if the representation (3.4)–(3.6) is used in (2.102) and (2.104), with the velocity in the Cartesian coordinates written as

$$\mathbf{u} = (u, v, w), \qquad (3.7)$$

the resulting vorticity equation is

$$\beta v = f \frac{\partial w}{\partial z}, \qquad (3.8)$$

which may be compared directly to (3.2). Thus little is lost dynamically, and much is gained in notational simplicity and analytical convenience, by considering the β-plane form of the planetary geostrophic equations, despite the geometrical distortion that is thereby incurred.

The Cartesian representation of the gradient operators $\nabla_h$ and $\nabla$, with (3.4)–(3.6), are the only modifications of (2.102)–(2.107) required to obtain the β-planc planetary geostrophic equations. For example, on the β-plane, the planetary geostrophic horizontal momentum equations (2.102) become

$$fv = \frac{1}{\rho_0} \frac{\partial p}{\partial x}, \quad fu = -\frac{1}{\rho_0} \frac{\partial p}{\partial y}, \tag{3.9}$$

and the material derivative operator is

$$\frac{D}{Dt} = \frac{\partial}{\partial t} + u \frac{\partial}{\partial x} + v \frac{\partial}{\partial y} + w \frac{\partial}{\partial z}. \tag{3.10}$$

### 3.3 Boundary Conditions on Vertical Velocity

Through the fundamental large-scale vorticity relation (3.2) or (3.8), the vertical velocity field, including top and bottom boundary conditions, assumes a central role in large-scale circulation dynamics. Suppose that the ocean is confined vertically between a sea floor at depth $z = -H(x, y)$ and a free surface at $z = \zeta(x, y, t)$. Conditions for the vertical velocity $w$ on these two boundaries can be obtained by the standard fluid mechanical requirement that fluid parcels initially on the boundary must remain there so that

$$\frac{D(z - \zeta)}{Dt} = \frac{D(z + H)}{Dt} = 0. \tag{3.11}$$

These imply

$$w = \frac{D\zeta}{Dt} \quad \text{at } z = \zeta(x, y, t) \tag{3.12}$$

and

$$w = -\mathbf{u}_h \cdot \nabla_h H \quad \text{at } z = -H(x, y). \tag{3.13}$$

For a characteristic free-surface deformation $\Delta\zeta = 1$ m and time or advection scales from (2.1), the rate of change of free-surface height can be estimated as

$$\left| \frac{D\zeta}{Dt} \right| \approx \frac{\Delta\zeta}{t_{adv}} \approx 2 \times 10^{-10} \text{ s}^{-1} \ll W = 10^{-5} \text{ s}^{-1}, \tag{3.14}$$

where $W$ is the characteristic magnitude of the planetary geostrophic velocity $w$. It follows that it should be appropriate also to replace (3.12) with the simplified surface boundary condition

$$w = 0 \quad \text{at } z = 0. \tag{3.15}$$

The condition (3.15) can be consistently imposed at $z = 0$ because $\Delta\zeta \ll D$, where $D$ is the characteristic vertical scale of the large-scale motion from (2.1).

## 3.4 The Transport Stream Function

The boundary conditions (3.13) and (3.15) on the vertical velocity, along with the mass conservation equation (2.104), imply that the vertically integrated horizontal flow $\mathbf{U}_S(x, y)$ is incompressible:

$$0 = \int_{-H(x,y)}^{0} \left( \nabla_h \cdot \mathbf{u}_h + \frac{\partial w}{\partial z} \right) dz = \nabla_h \cdot \mathbf{U}_S, \tag{3.16}$$

where

$$\mathbf{U}_S = \int_{-H(x,y)}^{0} \mathbf{u}_h \, dz. \tag{3.17}$$

This in turn means that a stream function $\Psi(x, y, t)$ can be defined for the vertically integrated flow so that

$$\mathbf{U}_S = (U_S, V_S) = \left( -\frac{\partial \Psi}{\partial y}, \frac{\partial \Psi}{\partial x} \right). \tag{3.18}$$

This result holds even if $H$ depends on $x$ and $y$ and does not depend on the geostrophic or hydrostatic balance or on the vorticity relation (3.8). It also does not depend on the $\beta$-plane approximation as an analogous stream function can be defined in spherical coordinates.

## 3.5 The Wind-Driven Surface Ekman Layer

Vertical motion can be forced in a turbulent boundary layer at the sea surface by the action of the wind stress on the ocean. The simplest theoretical representations of such turbulent boundary layers, in which the Coriolis force associated with horizontal motion in the boundary layer balances the vertical divergence of turbulent stress, are known as Ekman layers. The boundary condition (3.15) still applies to the total vertical velocity, but the boundary condition on the component of the vertical velocity that is associated with the geostrophic motion must be modified to compensate for the induced vertical motion in the ageostrophic boundary layer.

The horizontal momentum balance in the turbulent surface Ekman layer may be written as

$$f\mathbf{k} \times (\mathbf{u}_h + \mathbf{u}_E) = -\frac{1}{\rho_0} \nabla_h p - \frac{1}{\rho_0} \frac{\partial \boldsymbol{\tau}_E}{\partial z}, \tag{3.19}$$

where $\mathbf{u}_E = (u_E, v_E)$ is the horizontal Ekman velocity vector, representing the departure of the boundary-layer velocity from the geostrophic velocity $\mathbf{u}_h$, and the stress vector $\boldsymbol{\tau}_E = (\tau^x, \tau^y)$ is the horizontal component of the vertical turbulent flux of horizontal momentum. The stress divergence is analogous to the terms $(\nabla \cdot \mathbf{d})_h$ that were dropped from the momentum equation (1.16) in deriving the inviscid form (2.95). In Ekman layer theory, however, the kinematic stresses $\boldsymbol{\tau}_E$ are turbulent momentum fluxes, which represent the mean effect of fluctuating small-scale continuum fluid motions, rather than viscous momentum fluxes arising from collisions of molecules.

Because the geostrophic terms are linear, Equation (3.19) may be simplified to

$$f\mathbf{k} \times \mathbf{u}_E = -\frac{1}{\rho_0} \frac{\partial \boldsymbol{\tau}_E}{\partial z}. \tag{3.20}$$

Observations suggest a characteristic scale $\delta_E$ of order 100 m or less for the thickness of the turbulent surface boundary layer, in which the stress $\boldsymbol{\tau}_E$ is large. An analytical solution for the velocity profile in the Ekman layer is available for the case in which the turbulent dynamics are represented by a constant vertical eddy viscosity $A_V$. The stress vector in (3.20) is then $\boldsymbol{\tau}_E = -A_V \partial \mathbf{u}_E/\partial z$, and the resulting horizontal velocities decay exponentially with depth over a vertical scale given by the Ekman depth $(2A_V/f)^{1/2}$. To derive the required boundary condition on the large-scale vertical velocity, however, it is not necessary to consider the details of this solution. Instead, it is sufficient to assume that the velocity and stress vanish at depths much larger than the boundary layer depth $\delta_E$ and to consider the vertical integral over the boundary layer of the basic balance (3.20) of Coriolis force and stress divergence. This yields the vertically integrated horizontal boundary layer transport $\mathbf{U}_E$:

$$\mathbf{U}_E = (U_E, V_E) = \int_{z/\delta_E \to -\infty}^{0} \mathbf{u}_E \, dz = -\mathbf{k} \times \frac{\boldsymbol{\tau}_w}{\rho_0 f} = \frac{1}{\rho_0 f} \left( \tau_w^y, -\tau_w^x \right). \tag{3.21}$$

Here $\boldsymbol{\tau}_w = (\tau_w^x, \tau_w^y)$ is the surface wind stress, the flux of momentum from the atmosphere to the ocean, now defined to be positive downward. In (3.21), the lower integral limit represents a depth well below the nominal Ekman layer depth $z = -\delta_E$, and the upper integral limit is taken as $z = 0$ rather than $z = \zeta$, as in (3.15). Thus, remarkably, the horizontal Ekman layer transport depends only on the wind stress $\boldsymbol{\tau}_w$ and not on the details of the turbulence in the boundary layer.

The desired boundary condition for the vertical velocity of the geostrophic flow can be obtained from the convergence of the Ekman transport (3.21). Define an Ekman vertical velocity $w_E$ through an incompressibility condition of the form (2.38):

$$\frac{\partial w_E}{\partial z} = -\nabla_h \cdot \mathbf{u}_E. \tag{3.22}$$

Because $\mathbf{u}_E \to 0$ as $z/\delta_E \to -\infty$, it is natural to require also that $w_E \to 0$ as $z/\delta_E \to -\infty$ so that the Ekman flow is a boundary-layer correction, entirely confined to the boundary layer. The vertical integral of (3.22) then gives

$$w_E|_{z=0} = -\int_{-\infty}^{0} \nabla_h \cdot \mathbf{u}_E \, dz = -\nabla_h \cdot \mathbf{U}_E = -W_E, \tag{3.23}$$

where $W_E$ is defined by

$$W_E = \frac{1}{\rho_0} \left[ \frac{\partial}{\partial x} \left( \frac{\tau_w^y}{f} \right) - \frac{\partial}{\partial y} \left( \frac{\tau_w^x}{f} \right) \right]. \tag{3.24}$$

The no-normal-flow boundary condition (3.15) now must hold for the total vertical velocity at $z = 0$, the sum of the Ekman component $w_E$ and the planetary geostrophic component $w$:

$$w + w_E = 0 \quad \text{at } z = 0. \tag{3.25}$$

The resulting boundary condition on the planetary geostrophic vertical velocity at the sea surface $z = 0$ is

$$w|_{z=0} = W_E = \frac{1}{\rho_0} \left[ \frac{\partial}{\partial x} \left( \frac{\tau_w^y}{f} \right) - \frac{\partial}{\partial y} \left( \frac{\tau_w^x}{f} \right) \right]. \tag{3.26}$$

With characteristic scales of $\tau_* = 0.1$ N m$^{-2}$ for the wind stress and $\tau_*/L_\tau$ with $L_\tau = 10^6$ m for the curl of the wind stress, (3.26) gives a vertical velocity scale $W \approx W_E \approx \tau_*/(\rho_0 L_\tau f_0) \approx 10^{-6}$ m s$^{-1}$, as anticipated in (2.1). Note that for $\delta_E = 50$ m, the characteristic wind-driven horizontal velocity within the Ekman layer is $U_E/\delta_E = \tau_*/(\rho_0 f_0 \delta_E) \approx 2 \times 10^{-2}$ m s$^{-1} \gg U$, where $U$ is the horizontal velocity scale for the geostrophic flow from (2.1).

If there is evaporation or precipitation at the sea surface, then the original boundary condition (3.12) on the total vertical velocity must be modified to account for the corresponding rate $w_*$ of extraction or addition of fluid at the interface so that

$$w = \frac{D\zeta}{Dt} + w_* \quad \text{at } z = \zeta(x, y, t). \tag{3.27}$$

If $w_* \neq 0$, the total vertically integrated flow is no longer divergence free, and the stream function (3.18) for the total transport must be supplemented by a transport potential. Typical maximum values of $w_*$ for the ocean are of order 1 m yr$^{-1}$. This is substantially smaller than the typical values of midlatitude Ekman pumping $W_E$, which, from (2.1), are of order $W = 10^{-6}$ m s$^{-1} \approx 30$ m yr$^{-1}$. The circulations driven directly by evaporation and precipitation may therefore be ignored at first order.

## 3.6 Sverdrup Transport

The planetary geostrophic vorticity relation (3.8) is linear in both $v$ and $\partial w/\partial z$ and may be integrated over depth to obtain

$$\beta V_G = \beta \int_{-H}^{0} v \, dz = f(w|_{z=0} - w|_{z=-H}). \tag{3.28}$$

Here the approximation $\zeta \approx 0$ has been made in the upper integration limit because $\zeta \ll D$, where $D$ is a characteristic scale for $H$ from Section 2.1. Thus the vertically integrated meridional geostrophic transport $V_G$ can be computed directly from the boundary conditions on the vertical velocity.

If the sea floor is flat so that $H(x, y) = H_0$, where $H_0$ is a constant, or $H$ is large enough that the vertical motion vanishes at great depths, then (3.13) reduces immediately to one of the simple conditions

$$w = 0 \quad \text{at } z = -H_0 \quad \text{or} \quad w \to 0 \quad \text{as } z \to -\infty. \tag{3.29}$$

The result of imposing (3.29) and the original no-normal-flow condition (3.15) at $z = 0$ in the vertically integrated equation (3.28) is that $V_G = 0$: if the circulation is not forced by vertical motions at the boundaries, there can be no net vertically integrated meridional geostrophic flow at each point in the domain.

With, instead, the Ekman vertical velocity boundary condition (3.26) at $z = 0$, and either of (3.29), the vertical integral of the vorticity relation (3.8) yields

$$\beta V_G = \frac{f}{\rho_0} \left[ \frac{\partial}{\partial x} \left( \frac{\tau_w^y}{f} \right) - \frac{\partial}{\partial y} \left( \frac{\tau_w^x}{f} \right) \right]. \tag{3.30}$$

According to (3.30), the vertically integrated meridional geostrophic transport $V_G$ at each point is proportional to the curl of the ratio of the wind stress $\tau_w$ to the local Coriolis parameter $f$. A characteristic scale for $V_G$ is $V_G \approx \tau_*/(\rho_0 \beta L_\tau) \approx 5 \text{ m}^2 \text{ s}^{-1}$.

Equation (3.30) is the geostrophic Sverdrup transport balance, a fundamental element of large-scale circulation theory. It follows from the scaling (2.1) and from the result (3.26) of Ekman boundary layer theory, which does not depend on details of the representation of turbulence in the boundary layer. Through this balance, the vertically integrated meridional geostrophic transport can be computed directly from the surface wind stress without any consideration of any other aspects of the flow. In particular, note that (3.30) does not depend on any restriction on the density structure of the flow, provided that the flow is consistent with the a priori scales, and so holds for both homogeneous and stratified planetary geostrophic flow.

The term $\beta V_G$ in (3.30) is proportional to the divergent part of the geostrophic flow. This can be seen as follows. Consider the vertical integral of the geostrophic relations (3.9):

$$\mathbf{U}_G = (U_G, V_G) = \frac{1}{\rho_0 f} \int_{-H_0}^{0} \mathbf{k} \times \nabla_h p \, dz. \tag{3.31}$$

The divergence of the geostrophic flow is then

$$\nabla_h \cdot \mathbf{U}_G = -\frac{\beta}{\rho_0 f^2} \int_{-H_0}^{0} \frac{\partial p}{\partial x} \, dz = -\frac{\beta}{f} V_G. \tag{3.32}$$

Since the sum of the geostrophic and Ekman flows satisfies the boundary condition (3.25), the horizontal divergence of the sum of the vertically integrated geostrophic and Ekman transports must vanish, as can be verified directly from (3.32), (3.30), and (3.21):

$$\nabla_h \cdot (\mathbf{U}_G + \mathbf{U}_E) = -\frac{\beta}{f} V_G + \frac{\partial}{\partial x} \left( \frac{\tau_w^y}{\rho_0 f} \right) + \frac{\partial}{\partial y} \left( -\frac{\tau_w^x}{\rho_0 f} \right) = 0. \qquad (3.33)$$

This means that a stream function $\Psi(x, y, t)$ can still be defined for the total vertically integrated flow so that

$$\mathbf{U}_S = (U_S, V_S) = \mathbf{U}_G + \mathbf{U}_E = \left( -\frac{\partial \Psi}{\partial y}, \frac{\partial \Psi}{\partial x} \right). \qquad (3.34)$$

Then it follows also from (3.30) and (3.21) that the product of $\beta$ and the total meridional transport is

$$\beta \frac{\partial \Psi}{\partial x} = \beta V_G + \beta V_E = \frac{1}{\rho_0} \left( \frac{\partial \tau_w^y}{\partial x} - \frac{\partial \tau_w^x}{\partial y} \right). \qquad (3.35)$$

According to (3.35), the sum of the vertically integrated wind-driven Ekman layer and planetary geostrophic meridional transports is proportional, at each point and at any given time, to the local value of the curl of the wind stress. Equation (3.35) is the Sverdrup transport balance; like (3.30), it is a fundamental result. On the time scale $t_{adv}$ from (2.1), both are quasi-steady balances.

## 3.7 Depth-Integrated Wind-Driven Gyre Circulation

The Sverdrup transport balance (3.35) allows the computation of the zonal gradient $\partial \Psi / \partial x$ of the stream function $\Psi$ for the depth-integrated flow from an imposed wind stress field $\boldsymbol{\tau}_w(x, y, t)$. For inviscid flow in a closed, simply connected domain, the natural boundary condition on a stream function is that it be constant, or a function only of time, on all rigid boundaries. If the no-normal-flow condition $\Psi(x_E, y, t) = 0$ is imposed at the eastern boundary $x = x_E$, the westward zonal integral of (3.35) gives the depth-integrated interior circulation at each longitude $x$, latitude $y$, and time $t$:

$$\Psi(x, y, t) = \frac{1}{\beta} \int_{x_E}^{x} \frac{1}{\rho_0} \left( \frac{\partial \tau_w^y}{\partial x} - \frac{\partial \tau_w^x}{\partial y} \right) dx. \qquad (3.36)$$

Thus the full two-dimensional structure of the depth-integrated, wind-driven, large-scale interior circulation can be computed from knowledge only of the wind stress field. This is a fundamental result with wide-reaching implications. A characteristic scale for the transport $\Psi$ is $\Psi \approx V_G L \approx 2.5 \times 10^7$ m$^3$ s$^{-1}$ = 25 Sv, where 1 Sv (Sverdrup) = $10^6$ m$^3$ s$^{-1}$.

However, from a theoretical point of view, the choice made in (3.36) to impose the no-normal-flow condition at the eastern rather than western boundary appears arbitrary and even inconsistent. In a simple closed domain, the no-normal-flow condition must be specified along the entire boundary, and the value of $\Psi$ at the eastern and western boundaries should be the same. In (3.35), however, only one boundary condition can be specified independently, in the zonal coordinate $x$. A zonal integral of (3.35) across the basin from the western boundary $x_W$ to the eastern boundary $x_E$ gives

$$\beta(\Psi_E - \Psi_W) = \int_{x_W}^{x_E} \frac{1}{\rho_0} \left( \frac{\partial \tau_w^y}{\partial x} - \frac{\partial \tau_w^x}{\partial y} \right) dx, \tag{3.37}$$

where $\Psi_E = \Psi(x_E, y, t)$ and $\Psi_W = \Psi(x_W, y, t)$. Thus the two boundary conditions $\Psi_E = \Psi_W = 0$ can only be satisfied simultaneously if the zonal integral of the wind stress curl happens to vanish, which, in general, it will not. The source of this degeneracy is the neglect of higher-order spatial derivative operators in the derivation of the reduced equations (2.102)–(2.107), especially those associated with the molecular viscosity and the divergence of deviatoric stress $\nabla \cdot \mathbf{d}$.

To overcome this degeneracy, it is necessary to restore some representation of frictional effects in the dynamical equations. One approach is to introduce a frictional drag force that is linearly proportional to the flow and directed opposite to it. Then the horizontal momentum equations for the vertically integrated flow, written in terms of the transport stream function $\Psi$, are

$$-f \frac{\partial \Psi}{\partial x} = -\frac{\partial \bar{P}}{\partial x} + \frac{\tau_w^x}{\rho_0} + r \frac{\partial \Psi}{\partial y}, \quad -f \frac{\partial \Psi}{\partial y} = -\frac{\partial \bar{P}}{\partial y} + \frac{\tau_w^y}{\rho_0} - r \frac{\partial \Psi}{\partial x}, \tag{3.38}$$

where the last term in each of (3.38) is the drag force, $r$ is a constant drag coefficient, and

$$\bar{P} = \frac{1}{\rho_0} \int_{-H}^{0} p \, dz \tag{3.39}$$

is the vertically integrated pressure. The main motivation for introducing this particular representation of the neglected frictional effects is the simplicity of the equations (3.38) that follow from it rather than a specific physical argument.

A vorticity equation for the depth-integrated flow can be obtained by cross-differentiation of (3.38), which eliminates the pressure $\bar{P}$, with the result that

$$r \nabla_2^2 \Psi + \beta \frac{\partial \Psi}{\partial x} = \frac{1}{\rho_0} \left( \frac{\partial \tau_w^y}{\partial x} - \frac{\partial \tau_w^x}{\partial y} \right). \tag{3.40}$$

This equation is of second differential order in both $x$ and $y$ so that the natural condition $\Psi = 0$ can be consistently imposed along the boundary of a simple, closed domain.

It is instructive to consider solutions of (3.40) in a rectangular closed basin with boundaries at $x = \{x_W, x_E\}$ and $y = \{0, L\}$ for a wind stress field $\boldsymbol{\tau}_w = (\tau_w^x, \tau_w^y)$

given by

$$\tau_w^x = -\tau_0 \cos\left(\frac{\pi y}{L}\right) \quad \text{and} \quad \tau_w^y = 0, \tag{3.41}$$

where $\tau_0$ is a constant. The form (3.41) is a simple, zonally symmetric representation of the mean midlatitude wind stress, with easterlies at the lower latitudes ($0 < y < L/2$) and westerlies at the higher latitudes ($L/2 < y < L$). With the substitution

$$\Psi(x, y) = \Phi(x) \sin\left(\frac{\pi y}{L}\right), \tag{3.42}$$

for which $\Psi = 0$ on $y = \{0, L\}$ is automatically satisfied, Equation (3.40) becomes

$$r\frac{d^2\Phi}{dx^2} + \beta\frac{d\Phi}{dx} - r\frac{\pi^2}{L^2}\Phi = -\frac{\pi\tau_0}{\rho_0 L}. \tag{3.43}$$

The inviscid ($r = 0$) solution of (3.43) is

$$\Phi(x) = -\frac{\pi\tau_0}{\rho_0 \beta L}x + C_0, \tag{3.44}$$

where $C_0$ is a constant to be determined. Because a single value of $C_0$ cannot be chosen that will satisfy the boundary conditions $\Psi = 0$ at both of $x = \{x_W, x_E\}$, the presence of a frictional boundary layer adjacent to one of the boundaries can be anticipated. Suppose that this boundary layer has scale $l \ll L$, and introduce the boundary layer coordinate

$$\xi = (x - x_0)/l, \tag{3.45}$$

where $x_0$ represents either $x_W$ or $x_E$. With the substitution (3.45), (3.43) may be written as

$$\frac{r}{\beta l}\frac{d^2\Phi}{d\xi^2} + \frac{d\Phi}{d\xi} = \frac{l}{L}\left(\frac{r\pi^2}{\beta L}\Phi - \frac{\pi\tau_0}{\rho_0\beta}\right). \tag{3.46}$$

Now, choose

$$l = r/\beta; \tag{3.47}$$

let

$$\delta_S = l/L; \tag{3.48}$$

expand $\Phi$ in powers of $\delta_S$,

$$\Phi = \Phi_0 + \delta_S\Phi_1 + \delta_S^2\Phi_2 + \dots; \tag{3.49}$$

substitute these into (3.46); and collect the first-order terms in $\delta_S$ to obtain

$$\frac{d^2\Phi_0}{d\xi^2} + \frac{d\Phi_0}{d\xi} = 0. \tag{3.50}$$

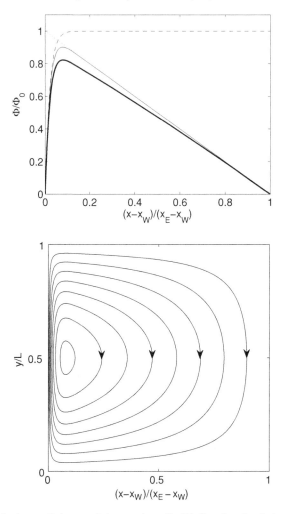

Figure 3.1. Solutions of the model equation (3.43) for the depth-integrated wind-driven gyre circulation for $\delta_S = 0.02$. (top) Numerical solution (Equation (3.40); thick solid line) and interior (Equation (3.44); dotted line), boundary layer (Equation (3.51); dashed line), and uniformly valid (Equation (3.52), thin solid line) analytical approximations. (bottom) $\Phi/\Phi_0$ (contour interval 0.1; contours increase monotonically from boundary) from numerical solution of (3.40), using (3.42) and the numerical solution of (3.43).

Equation (3.50) has solution

$$\Phi_0 = C_1 + C_2 e^{-\xi} = C_1 + C_2 e^{-\beta(x-x_0)/r}, \tag{3.51}$$

where $C_1$ and $C_2$ are constants. Because the exponential in (3.51) decays as $\beta x/r \to +\infty$ but grows as $\beta x/r \to -\infty$, the bounded solution must have $x_0 = x_W$, and the boundary layer must be adjacent to the western boundary (Figure 3.1). The corresponding uniformly valid approximate solution constructed from (3.44) and

(3.51), accurate to first order in $\delta_S$, is

$$\Psi(x, y) = \frac{\pi \tau_0}{\rho_0 \beta L} \left[ (x_E - x) - (x_E - x_W)e^{-\beta(x-x_W)/r} \right] \sin \left( \frac{\pi y}{L} \right). \qquad (3.52)$$

For $x - x_W \gg r/\beta$, that is, outside the narrow western boundary layer, the solution (3.52) is equal to the inviscid interior solution (3.44), with uniform equatorward flow proportional to $\tau_0$ at each latitude $y$. The interior solution satisfies the eastern boundary condition $\Psi = 0$ on $x = x_E$, so there is no eastern boundary layer. This result is consistent with the choice made to impose the boundary condition on $\Psi$ at the eastern boundary in the general integral (3.36).

For $x - x_W \approx r/\beta$, that is, inside the western boundary layer, the solution (3.52) adjusts exponentially from its westernmost interior value, proportional to $x_E - x_W$, to the boundary value $\Psi = 0$ at $x = x_W$. In this narrow western boundary layer, $V_S = \Psi_x$ is large, and there is fast poleward flow, which balances the slow equatorward flow in the interior. If the value of $\beta$ in (3.52) is decreased toward zero, the width $r/\beta$ of the western boundary layer increases. If $\beta = 0$, then (3.40) and (3.43) are invariant under the transformation $x \to -x$, and the corresponding solution for $\Psi$ is symmetric with respect to $x$. Therefore it is the $\beta$-effect, the variation of the Coriolis parameter with latitude, that causes the western intensification of the flow, that is, the development of the narrow, fast return current along the western boundary of the domain.

The vorticity dynamics of the solution (3.52) are linear. The Sverdrup transport balance (3.35) obtains in the interior regime, with the anticyclonic wind stress acting to reduce the vorticity of the interior flow, and the interior fluid moving equatorward to match its vorticity with the ambient, planetary vorticity, following the planetary geostrophic vorticity balance. In the western boundary current, the meridional flow is increasingly faster—and consequently, the retarding force is increasingly greater— as the boundary is approached. This force differential imparts cyclonic vorticity to the fluid in the return current, allowing it to match the planetary vorticity at the corresponding latitude as it leaves the western boundary layer and returns to the Sverdrup interior. A similar frictional boundary layer could not have this effect along the eastern boundary as the greater drag adjacent to the boundary would impart the wrong sign of vorticity to poleward-flowing fluid.

Thus, in addition to allowing a consistent solution for the depth-integrated circulation in a closed basin that exhibits the planetary geostrophic vorticity balance in the inviscid interior, the simple drag parameterization in (3.38) leads to an appealing explanation for the origin of observed midlatitude western boundary currents such as the Gulf Stream in the North Atlantic. A limitation of this calculation is that the linear drag terms, through which momentum is removed from the depth-integrated flow, are not motivated by any quantitative physical considerations. An alternative model can be obtained by specifying lateral, down-gradient eddy diffusion of momentum in place of the linear drag. This would allow the flow to satisfy a full no-slip condition along the lateral boundaries, but also is not directly motivated by quantitative physical considerations. In the special case of a uniform-density fluid, for which the geostrophic flow is independent of depth, the linear drag terms can be interpreted physically as a parameterized bottom friction; this interpretation is directly relevant to laboratory experiments but not to the stratified, large-scale ocean gyre circulations.

### 3.8 Topographic Effects on the Depth-Integrated Flow

If the flow extends to the seafloor and the seafloor is not flat, then the lower boundary condition (3.13) on vertical velocity must be retained in its full form. Through the vertical integral (3.28) of the planetary geostrophic vorticity balance (3.8), this couples the vertically integrated meridional transport to the horizontal velocity $\mathbf{u}_h$ at $z = -H(x, y)$:

$$\beta \frac{\partial \Psi}{\partial x} = \frac{1}{\rho_0} \left( \frac{\partial \tau_w^y}{\partial x} - \frac{\partial \tau_w^x}{\partial y} \right) + f \mathbf{u}_h|_{z=-H} \cdot \nabla_h H. \tag{3.53}$$

According to (3.53), the depth-integrated rate of change of planetary vorticity is equal to the difference of the vortex stretching terms arising from Ekman pumping at the surface and from the vertical velocity at the bottom, where the latter is in turn forced by bottom geostrophic flow up the topographic gradient.

Alternatively, with the geostrophic relations (3.9), by which the horizontal velocity $\mathbf{u}_h$ at $z = -H(x, y)$ is related to the bottom pressure $p_b(x, y, t) = p[x, y, -H(x, y), t]$, the modified, topographic Sverdrup transport balance (3.53) can also be written as

$$\beta \frac{\partial \Psi}{\partial x} = \frac{1}{\rho_0} J(p_b, H) + \frac{1}{\rho_0} \left( \frac{\partial \tau_w^y}{\partial x} - \frac{\partial \tau_w^x}{\partial y} \right), \tag{3.54}$$

where the Jacobian $J(a, b)$ of two functions $a(x, y)$ and $b(x, y)$ is defined by

$$J(a, b) = \frac{\partial a}{\partial x} \frac{\partial b}{\partial y} - \frac{\partial a}{\partial y} \frac{\partial b}{\partial x}. \tag{3.55}$$

The term $J(p_b, H)$ is the bottom pressure torque. It vanishes only when the gradients of bottom pressure and topography are parallel or if at least one of those two gradients vanishes. The equation (3.54) may also be obtained by writing the vertically integrated momentum equations in the form

$$-f \Psi_x = -\frac{1}{\rho_0} \int_{-H(x,y)}^{0} \frac{\partial p}{\partial x} dz + \frac{\tau_w^x}{\rho_0} \tag{3.56}$$

$$-f \Psi_y = -\frac{1}{\rho_0} \int_{-H(x,y)}^{0} \frac{\partial p}{\partial y} dz + \frac{\tau_w^y}{\rho_0}, \tag{3.57}$$

where $\Psi$ is as defined in (3.18), and then cross-differentiating, taking account of the variable lower limit of integration $z = -H(x, y)$.

A related vorticity equation can be obtained from the vertically averaged momentum equations, as follows. The pressure term in (3.56) may be integrated by parts:

$$\int_{-H}^{0} p \, dz = \int_{-Hp_b}^{0} d(zp) - \int_{p_b}^{p_a} z \, dp = Hp_b + \mathcal{P}, \qquad (3.58)$$

where

$$\mathcal{P} = g \int_{-H}^{0} z\rho \, dz. \qquad (3.59)$$

If (3.58) is used in (3.56)–(3.57) after manipulation of the pressure-gradient integrals, and the result is divided by $H$ and then cross-differentiated, the resulting equation is

$$J\left(\Psi, \frac{f}{H}\right) - J\left(\frac{\mathcal{P}}{\rho_0}, \frac{1}{H}\right) = \frac{1}{\rho_0}\left[\frac{\partial}{\partial x}\left(\frac{\tau_w^y}{H}\right) - \frac{\partial}{\partial y}\left(\frac{\tau_w^x}{H}\right)\right]. \qquad (3.60)$$

In the special case of a fluid with uniform density $\rho = \rho_0$, it follows that $\mathcal{P} = -\rho_0 g H^2/2$, and thus $J(\mathcal{P}, 1/H) = 0$ identically. In this case, Equation (3.60) shows that the total depth-integrated flow crosses contours of constant $f/H$ only to the extent that it is forced by the curl of the ratio of the wind stress $\boldsymbol{\tau}_w$ to the bottom depth $H$.

### 3.9 Notes

The Sverdrup vorticity relation and the Sverdrup transport balance were derived by Sverdrup (1947). The linear-frictional model of the depth-integrated wind-driven circulation was constructed by Stommel (1948) as the first explanation of the observed western intensification of the midlatitude gyres; an extension to Laplacian horizontal momentum diffusion was developed by Munk (1950). Note that these analyses of the large-scale vorticity balance and its role in the depth-integrated basin-scale flow preceded the first derivation of the full planetary geostrophic equations by a decade. A detailed discussion of Ekman layer theory, including the depth-dependent structure within the boundary layer, is given by Pedlosky (1987).

# 4

# Stratified Large-Scale Flow

## 4.1 Planetary Geostrophic Equations for a Simplified Equation of State

The vorticity dynamics explored in the previous chapter provides powerful constraints on the depth-integrated planetary geostrophic flow. However, it provides no information on how the horizontal flow is distributed in the vertical, nor on the interaction of dynamic and thermodynamic processes that set the large-scale thermal and haline structure of the ocean. These elements depend on the stratification of the ocean—the variation of density with depth—and the dependence of density on the three-dimensional thermal and haline fields, through the equation of state.

The basic equation of state for seawater is the nonlinear, empirical relation (1.14):

$$\rho = \mathcal{R}(p, T, S). \tag{4.1}$$

As a starting point for the stratified theory, it is sufficient to use an approximate, linear form of (4.1):

$$\rho = \rho_0 \left\{ 1 + \Gamma_\rho [\bar{p}(z) + p' - p_r] - \alpha_T (T - T_r) + \beta_S (S - S_r) \right\}, \tag{4.2}$$

where the compressibility $\Gamma_\rho$, the thermal expansion coefficient $\alpha_T$, and the haline contraction coefficient $\beta_S$ are taken to be constant. In (4.2), the total pressure $p$ has been written as the sum of a horizontally uniform component $\bar{p}(z)$ and a departure $p'(x, y, z, t)$ from $\bar{p}(z)$, where $\bar{p}(z)$ is taken to satisfy

$$\frac{d\bar{p}}{dz} = -g\rho_0 \left\{ 1 + \Gamma_\rho [\bar{p}(z) - p_r] + \alpha_T T_r - \beta_S S_r \right\}, \quad \bar{p}(z_r) = p_r \tag{4.3}$$

for a suitable reference level $z_r$ and constants $\rho_0$, $p_r$, $T_r$, and $S_r$ such that $\rho = \rho_0$ at $z = z_r$ when $p' = 0$, $T = T_r$, and $S = S_r$; here $\rho_0$ need not take the precise value given in (2.1). In this linear approximation to (4.1), in which the compressibility, expansion, and contraction coefficients are constants, the differential equation (4.3) can be solved for $\bar{p}(z)$, yielding

$$\bar{p}(z) = p_r + (1 + \alpha_T T_r - \beta_S S_r)\Gamma_\rho^{-1}\{\exp[-g\rho_0\Gamma_\rho(z - z_r)] - 1\}, \qquad (4.4)$$

or

$$\bar{p}(z) \approx p_r - g\rho_0(1 + \alpha_T T_r - \beta_S S_r)(z - z_r). \qquad (4.5)$$

This solution for $\bar{p}(z)$ is a trivial modification of (2.29), and the decomposition $p = \bar{p} + p'$ follows (2.33).

Because the horizontal gradient of the horizontally uniform pressure $\bar{p}(z)$ vanishes identically, and the equation of state (4.2) is linear, $\bar{p}(z)$ can be completely removed from the equations. The resulting hydrostatic equation for the departure $p'$ of the pressure from $\bar{p}(z)$ is then

$$\frac{\partial p'}{\partial z} = -g\rho_0(\Gamma_\rho p - \alpha_T T + \beta_S S) \approx g\rho_0(\alpha_T T - \beta_S S), \qquad (4.6)$$

where $\rho' = -\rho_0 (\alpha_T T - \beta_S S)$ is the density anomaly, and the last approximation follows because the neglected term is smaller than $\partial p'/\partial z$ by a factor of $\delta_\rho \approx g\rho_0\Gamma_\rho D \ll 1$. Thus it is only the linear combination $\alpha_T T - \beta_S S$ that has a dynamical effect on density and pressure. Because the evolution equations (2.105) and (2.106) are also linear in $T$ and $S$, one of these variables is dynamically redundant, provided that the boundary conditions may also be written in terms of $\alpha_T T - \beta_S S$. In this case, only one of $T$ and $S$ need be retained; equivalently, a new single variable proportional to $\alpha_T T - \beta_S S$ can be defined such as the buoyancy $b = g\rho_0(\alpha_T T - \beta_S S)$.

Under these conditions, there is no loss of generality in setting $S = 0$, which gives a simplified planetary geostrophic theory with only the single thermodynamic variable $T$. The result is a simplified set of planetary geostrophic equations for large-scale ocean circulation:

$$f\,\mathbf{k} \times \mathbf{u}_h = -\frac{1}{\rho_0}\nabla_h\, p', \qquad (4.7)$$

$$\frac{\partial p'}{\partial z} = g\,\rho_0\,\alpha_T\, T, \qquad (4.8)$$

$$\nabla \cdot \mathbf{u} = 0, \qquad (4.9)$$

$$\frac{DT}{Dt} = 0. \qquad (4.10)$$

Note that the density $\rho$ and the explicit equation of state have been eliminated from these equations, which are written entirely in terms of $T$. The horizontally uniform pressure field $\bar{p}(z)$ need only be considered explicitly if the in situ density is to be reconstructed from (4.2) and the solution for $T$.

The simplification afforded by the linearization of the equation of state, and the assumption either that $S = 0$ or that the combined boundary conditions can be consistently posed in terms of the combination $\alpha_T T - \beta_S S$, is substantial. The number of evolution equations is reduced from two to one, and a separate equation of state is no longer required to close the system. If temperature variations are sufficiently small that the potential temperature $\theta$ should be considered in place of the in situ temperature $T$, a set of equations results that has identical form to (4.2)–(4.10), with $T$ simply replaced everywhere by $\theta$.

The system (4.7)–(4.10) is traditionally known also as the *thermocline equations*. It is frequently written in terms of the density anomaly $\rho'$ instead of temperature $T$. However, that formulation is purposely avoided here because the associated replacement of (4.10) with $D\rho'/Dt = 0$ obscures the origin of (4.10) as a form of the thermodynamic energy equation. The system may also be written in terms of the buoyancy variable $b$, with (4.10) replaced by $Db/Dt = 0$.

## 4.2 Linear Theory

Although the equation of state (4.2) has been linearized, the planetary geostrophic equations (4.7)–(4.10) still contain the fundamental advective nonlinearity: the velocity field that advects the temperature $T$ itself depends on $T$ through the hydrostatic and geostrophic balances. If it is assumed that the deviations from a stratified rest state are small, a linear theory may be derived, as follows.

Suppose that the temperature field can be decomposed into a known part $\hat{T}$ that varies only in the vertical plus a small disturbance $\tilde{T}$:

$$T(x, y, z, t) = \hat{T}(z) + \tilde{T}(x, y, z, t), \tag{4.11}$$

where the magnitude of $\tilde{T}$ is assumed small relative to that of $\bar{T}$. The pressure disturbance field $p'$ can then be decomposed similarly:

$$p'(x, y, z, t) = \hat{p}(z) + \tilde{p}(x, y, z, t), \tag{4.12}$$

where

$$\frac{d\hat{p}}{dz} = g\,\rho_0\,\alpha_T\,\hat{T} \quad \text{and} \quad \frac{\partial \tilde{p}}{\partial z} = g\,\rho_0\,\alpha_T\,\tilde{T}. \tag{4.13}$$

Because the horizontal velocities depend only on $\tilde{p}$, according to

$$f\,\mathbf{k} \times \mathbf{u}_h = -\frac{1}{\rho_0}\nabla_h \tilde{p}, \tag{4.14}$$

the horizontal advection terms in (4.10) are then quadratic in small disturbance quantities and may be neglected relative to the linear vertical advection and local time derivative terms. Thus (4.10) may be approximated by a linear equation:

$$\frac{\partial \tilde{T}}{\partial t} + w \frac{d\hat{T}}{dz} = \frac{\partial \tilde{T}}{\partial t} + \frac{N^2}{g\alpha_T} w = 0. \tag{4.15}$$

In (4.15), the vertical velocity $w$ is determined from (4.9), the form of which is unchanged in the linear theory, and

$$N^2(z) = g\alpha_T \frac{d\hat{T}}{dz} \tag{4.16}$$

is a simplified form of the buoyancy frequency (2.74).

## 4.3 Breakdown of Steady Linear Theory

The linearized thermodynamic energy equation (4.15), while apparently innocuous, hides an imminent catastrophe. Steady solutions, for which $\partial \tilde{T}/\partial t = 0$, must clearly have $w = 0$ everywhere in the domain. According to the Sverdrup vorticity relation (3.8), then, it follows also that $v = 0$ everywhere in the domain. This, however, is in direct contradiction to the geostrophic Sverdrup transport balance (3.30), according to which there must be a vertically integrated meridional geostrophic flow proportional to the curl of the ratio of the wind stress to the Coriolis parameter.

Two possible sources of this breakdown of the steady linear theory are the linearization itself, with its neglect of horizontal advection of thermal perturbations, and the restriction to purely large-scale physical processes, which has resulted in the purely adiabatic, advective form of the thermodynamic equation (4.10). The first of these can be explored within the context of the large-scale theory but inevitably requires the consideration of nonlinear equations. The second requires the inclusion of some representation of the integrated effect of small-scale processes on the large-scale flow. Both possibilities prove to be relevant to the large-scale circulation.

Remarkably, the time-dependent linear theory does not suffer the same catastrophe as the steady linear theory and instead yields useful insights into the mechanisms by which the large-scale circulation adjusts to changes in boundary forcing. Before proceeding with the steady theory, it is thus useful to consider the time-dependent solutions of the linearized equations.

## 4.4 Long Planetary Waves

For the linearized planetary geostrophic equations, an evolution equation in the single pressure variable $\tilde{p}$ can be derived by solving (4.15) and (4.13) for $w$,

$$w = -\frac{g\alpha_T}{N^2} \frac{\partial \tilde{T}}{\partial t} = -\frac{1}{\rho_0 N^2} \frac{\partial^2 \tilde{p}}{\partial z \partial t}, \tag{4.17}$$

and (4.14) for $v$,

$$v = \frac{1}{\rho_0 f} \frac{\partial \tilde{p}}{\partial x}, \tag{4.18}$$

and by substituting these into the Sverdrup vorticity relation (3.8). The result is conveniently written in the form

$$\frac{\partial}{\partial t} \left[ \frac{\partial}{\partial z} \left( \frac{f^2}{N^2} \frac{\partial \tilde{p}}{\partial z} \right) \right] + \beta \frac{\partial \tilde{p}}{\partial x} = 0. \tag{4.19}$$

In (4.19), the first term represents vortex stretching, $-f \, \partial w / \partial z$, and the second term represents the advection of ambient (planetary) vorticity, $\beta v$.

The pressure evolution equation (4.19) can be solved by separation of variables. Let

$$\tilde{p}(x, y, z, t) = P(x, t; y) \, Z(z; y), \tag{4.20}$$

where the dependence of $Z$ and $P$ on $y$ will only be parametric, through the dependence of $f$ on $y$. Then it follows that

$$\frac{1}{Z} \frac{d}{dz} \left( \frac{f^2}{N^2} \frac{dZ}{dz} \right) = -\frac{1}{\lambda^2} = -\beta \frac{\partial P}{\partial x} \left( \frac{\partial P}{\partial t} \right)^{-1}, \tag{4.21}$$

where $\lambda^2$ is a separation constant that depends on the parameter $y$.

The first equality in (4.21) gives an ordinary differential equation for the vertical structure function $Z$:

$$\frac{d}{dz} \left( \frac{f^2}{N^2} \frac{dZ}{dz} \right) + \frac{1}{\lambda^2} Z = 0. \tag{4.22}$$

For an ocean with a constant depth $H_0$, the natural boundary conditions for the free wave solutions of (4.19) are

$$w = 0 \quad \text{at } z = \{0, -H_0\}. \tag{4.23}$$

In view of (4.17) and (4.20), the corresponding boundary conditions on the vertical structure function $Z$ are

$$\frac{dZ}{dz} = 0 \quad \text{at } z = \{0, -H_0\}. \tag{4.24}$$

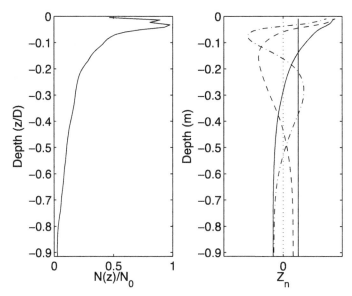

Figure 4.1. (left) Dimensionless long-term mean observed buoyancy frequency $N(z)/N_0$ vs. depth $z/D$ at 125°W, 30°N, for $N_0 = 10^{-2}$ s$^{-1}$ and $D = 5000$ m. (right) Corresponding normal modes $Z_j(z)$, $j = \{0, 1, 2, 3\}$, from Equations (4.22) and (4.24), vs. depth $z/D$, mode $j$ has $j$ zero crossings.

The second equality in (4.21) gives a first-order wave equation in $x$ and $t$:

$$\frac{\partial P}{\partial t} + c \frac{\partial P}{\partial x} = 0, \quad c = -\beta \lambda^2. \tag{4.25}$$

This has the general solution

$$P(x, t) = F(\xi), \quad \xi = x - ct, \tag{4.26}$$

where $F(\xi)$ is a smooth function, corresponding to a wave of fixed form that propagates with phase velocity $c$.

Equation (4.22) with boundary conditions (4.24), and with $f^2/N^2 > 0$ everywhere, forms an eigenvalue problem of Sturm-Liouville type, which has an infinite set of discrete real eigenvalues $\lambda_n^{-2}$, $n = \{0, 1, 2, \ldots\}$, $\lambda_{n+1}^{-2} > \lambda_n^{-2}$, with corresponding eigenfunctions, or vertical modes, $Z_n$ (Figure 4.1). Multiplying (4.22) by $Z$, integrating over $z$, and using (4.24) shows that

$$\frac{1}{\lambda^2} = \left( \int_{-H}^0 Z^2 \, dz \right)^{-1} \int_{-H}^0 \frac{f^2}{N^2} \left( \frac{dZ}{dz} \right)^2 \, dz > 0, \tag{4.27}$$

so the eigenvalues $\lambda_n^{-2}$ are all nonnegative. The differential operator in (4.22) is self-adjoint, so multiplying (4.22) for $Z_n$ by a different eigenfunction $Z_m$ and integrating over $z$ yields

$$\left(\frac{1}{\lambda_m^2} - \frac{1}{\lambda_n^2}\right) \int_{-H}^{0} Z_m Z_n \, dz = 0. \tag{4.28}$$

Thus distinct $(m \neq n)$ eigenfunctions are orthogonal with respect to the vertical integral.

By (4.27), the eigenvalues $\lambda_n^{-2}$ have dimensions of inverse length squared so that $\lambda_n$ has dimensions of length. For $n > 0$, the quantity $\lambda_n$ is the $n$th internal, or baroclinic, deformation radius. From (4.25), the corresponding wave speed $c_n$ of the $n$th vertical mode is

$$c_n = -\beta \lambda_n^2. \tag{4.29}$$

Because $c_n < 0$ for all $n > 0$, these waves all propagate toward negative $x$, that is, westward. These waves are the *baroclinic*, or internal, long planetary waves.

The $n = 0$ mode is the *barotropic*, or external, long planetary wave, and $\lambda_0$ is the barotropic, or external, deformation radius. With the simplified boundary condition $w = 0$ at $z = 0$ (4.23), this mode is degenerate: the solution is $Z_0 = \text{constant}$, $\lambda_0^{-2} = 0$, and $(\lambda_0, c_0) \to \infty$. A finite value for $\lambda_0$ can be obtained using a more accurate boundary condition for $w$ at $z = 0$ that accounts for the motion of the free surface. For a free surface at $z = \zeta(x, y, t)$, the linearized hydrostatic pressure fluctuation at $z = 0$ is, to first order in density, $\tilde{p} = g\rho_0\zeta$. Thus

$$w = \frac{1}{g\rho_0} \frac{\partial \tilde{p}}{\partial t} \quad \text{at } z = 0. \tag{4.30}$$

With (4.17) and (4.16), this yields

$$\frac{\partial \tilde{p}}{\partial z} = -\frac{N^2}{g} \tilde{p} \quad \text{at } z = 0 \tag{4.31}$$

so that the improved upper boundary condition for the barotropic vertical mode $Z_0(z)$ is

$$\frac{dZ_0}{dz} = -\frac{N^2}{g} Z_0 \quad \text{at } z = 0. \tag{4.32}$$

The condition (4.32) is sufficient to resolve the degeneracy, giving finite values for $\lambda_0$ and $c_0$, as in the example given in Section 4.5.

The long baroclinic and barotropic planetary waves are nondispersive: for any zonal wavenumber $k$, the corresponding Fourier mode for vertical mode $n$, $n = \{0, 1, 2, \ldots\}$,

$$P_{nk}(x, t) = \hat{P}_{nk} \, e^{-ik(x - c_n t)} = \hat{P}_{nk} \, e^{-i(kx - \omega_n t)}, \tag{4.33}$$

propagates at fixed phase velocity $c_n$, independent of $k$, and has frequency $\omega_n = kc_n$. Consequently, the group velocity $d\omega_n/dk = c_n$, and energy will propagate westward at the same speed as phase. In contrast, planetary waves arising in quasi-geostrophic theory, which is generally appropriate for motions with horizontal scales comparable to the deformation radius and much smaller than ocean basin widths, are dispersive, with frequency decreasing toward zero for wavenumbers larger than the inverse of the corresponding deformation radius. The planetary geostrophic approximation removes the relative vorticity that causes this quasi-geostrophic dispersion.

The long planetary waves obtained here from the planetary geostrophic equations are essentially equivalent to the long-wave limit ($k\lambda_n \ll 1$) of quasi-geostrophic planetary waves. The main distinction is that these long planetary waves have a parametric meridional dependence that is not present in the quasi-geostrophic representation, in which, for fixed central latitude $\phi_0$, the meridional structure of the waves may be sinusoidal. For both the baroclinic and the barotropic planetary geostrophic waves, the parametric dependence on the meridional coordinate $y$ enters through the effect of the variations of $f$ with $y$ on the structure of the vertical modes $Z_n$ and on the eigenvalues $\lambda_n$ and $c_n$.

These long planetary waves are the mechanism by which, in the linear approximation, the large-scale circulation responds to large-scale, low frequency changes in boundary forcing. The westward intensification that was discussed in the context of a steady, frictional model in Section 3.7 can be understood alternatively as arising in part from the uniformly westward energy propagation of these long waves.

## 4.5 Vertical Modes for Constant Stratification

Suppose that the stratification is constant so that

$$N^2(z) = N_0^2, \tag{4.34}$$

where $N_0$ is a constant buoyancy frequency. Then Equation (4.22) for the vertical modes becomes

$$\frac{d^2 Z}{dz^2} + \frac{N_0^2}{\lambda^2 f^2} Z = 0. \tag{4.35}$$

With the boundary conditions (4.24), the internal deformation radii $\lambda_n$ are

$$\lambda_n = \frac{N_0 H_0}{n\pi f}, \quad n = \{1, 2, 3, \ldots\}, \tag{4.36}$$

and the corresponding vertical modes $Z_n$ and phase velocities $c_n$ are

$$Z_n(z) = Z_{n0} \cos\left(\frac{n\pi z}{H_0}\right), \quad n = \{1, 2, 3, \ldots\} \tag{4.37}$$

$$c_n = -\beta\lambda_n^2 = -\beta\left(\frac{N_0 H_0}{n\pi f}\right)^2, \quad n = \{1, 2, 3, \ldots\}. \tag{4.38}$$

Thus the $n$th vertical mode $Z_n$ has $n$ zero crossings in $z$, and the internal deformation radii $\lambda_n$ and the magnitude of the phase velocities $|c_n|$ decrease as the inverse and inverse square, respectively, of the mode number $n$.

With the improved upper boundary condition (4.32), the barotropic vertical mode $Z_0$ may be approximated as

$$Z_0(z) = Z_{00}\left[1 - \frac{N_0^2}{2gH_0}(z + H_0)^2\right], \tag{4.39}$$

which, on substitution into (4.35) and neglect of small terms, yields the external deformation radius $\lambda_0$,

$$\lambda_0 = \frac{(gH_0)^{1/2}}{f}, \tag{4.40}$$

and the barotropic phase velocity $c_0$,

$$c_0 = -\beta\lambda_0^2 = -\beta\frac{gH_0}{f^2}. \tag{4.41}$$

Note that $c_1/c_0 = \lambda_1^2/\lambda_0^2 \approx \Delta\rho/\rho_0$, so the barotropic mode propagates about $10^3$ times faster than the first baroclinic mode. The polynomial form (4.39) can be obtained as an approximation to the sinusoidal solutions of (4.35) with (4.32), under the condition that $\lambda_0 \gg \lambda_1$, which is generally satisfied because $\lambda_1^2/\lambda_0^2 \approx \Delta\rho/\rho_0$.

Dimensional estimates of $\lambda_n$ and $c_n$ can be obtained using $N_0 = \{2 - 5\} \times 10^{-3}$ s$^{-1} \approx \{1 - 3\}$ cph and $H_0 = D = 5 \times 10^3$ m as characteristic values for the buoyancy frequency and ocean depth, respectively, with the midlatitude values $f = f_0 = 10^{-4}$ s$^{-1}$ and $\beta = 2 \times 10^{-11}$ m$^{-1}$ s$^{-1}$. The first internal deformation radius is then

$$\lambda_1 = \frac{N_0 H_0}{\pi f} \approx \{30 - 100\}\text{ km}, \tag{4.42}$$

and the phase velocity of the first baroclinic mode is

$$c_1 = -\beta\lambda_1^2 \approx -\{0.02 - 0.20\}\text{ m s}^{-1}. \tag{4.43}$$

Thus the first baroclinic mode will take several years to cross an ocean basin, and by (4.38) the crossing will take the $n$th baroclinic mode $n^2$ times as long as the first mode. For the barotropic mode,

$$\lambda_0 = \frac{(gH_0)^{1/2}}{f} \approx 2000\text{ km} \tag{4.44}$$

$$c_0 = -\beta\lambda_0^2 \approx -100\text{ m s}^{-1}. \tag{4.45}$$

Thus the barotropic mode will take only a day or so to cross an ocean basin at midlatitude. These basin-crossing times are useful as rough estimates for barotropic and first-mode baroclinic gyre adjustment tune scales. The rapid $n^{-2}$ decrease in $c_n$ with mode number $n$ makes the linearization about a state of rest increasingly unreliable for $n > 1$, since the phase speeds $c_n$ for $n \geq 2$ are generally comparable to or less than typical large-scale mean flow speeds.

## 4.6 Planetary Geostrophic Conservation Laws

The nonlinear planetary geostrophic equations (4.7)–(4.10) have two additional scalar conservation laws, along with (4.10) for temperature $T$. The first is for the planetary geostrophic potential vorticity $Q$, where

$$Q = f \frac{\partial T}{\partial z}. \tag{4.46}$$

Because

$$\frac{DQ}{Dt} = \frac{D}{Dt}\left(f \frac{\partial T}{\partial z}\right)$$

$$= f \frac{\partial}{\partial z}\left(\frac{\partial T}{\partial t} + u\frac{\partial T}{\partial x} + v\frac{\partial T}{\partial y} + w\frac{\partial T}{\partial z}\right)$$

$$- f\left(\frac{\partial u}{\partial z}\frac{\partial T}{\partial x} + \frac{\partial v}{\partial z}\frac{\partial T}{\partial y}\right) + \left(\beta v - f\frac{\partial w}{\partial z}\right)\frac{\partial T}{\partial z}, \tag{4.47}$$

it follows from the thermodynamic equation (4.10), the vorticity relation (3.8), and the thermal wind equations

$$\frac{\partial u}{\partial z} = -\frac{g\alpha_T}{f}\frac{\partial T}{\partial y}, \quad \frac{\partial v}{\partial z} = \frac{g\alpha_T}{f}\frac{\partial T}{\partial x} \tag{4.48}$$

that the potential vorticity $Q$ is conserved following the motion

$$\frac{DQ}{Dt} = 0. \tag{4.49}$$

This holds even for time-dependent flow, in which $\partial Q/\partial t$ and $\partial T/\partial t$ may be different from zero.

The second conservation law is for the Bernoulli function $B$ in steady flow, where

$$B = \frac{p'}{\rho_0} - g\,\alpha_T\,z\,T. \tag{4.50}$$

Because, with (4.10),

$$\frac{DB}{Dt} = \frac{1}{\rho_0}\frac{Dp'}{Dt} - g\alpha_T\left(wT + z\frac{DT}{Dt}\right)$$

$$= \frac{1}{\rho_0}\frac{\partial p'}{\partial t} - \frac{1}{\rho_0^2 f}\frac{\partial p'}{\partial y}\frac{\partial p'}{\partial x} + \frac{1}{\rho_0^2 f}\frac{\partial p'}{\partial x}\frac{\partial p'}{\partial y} + w\left(\frac{1}{\rho_0}\frac{\partial p'}{\partial z} - g\alpha_T T\right)$$

$$= \frac{1}{\rho_0}\frac{\partial p'}{\partial t}, \qquad\qquad (4.51)$$

it follows that under conditions of steady flow, in which $\partial B/\partial t = \partial p'/\partial t = 0$, the Bernoulli function $B$ is conserved following the steady motion

$$\mathbf{u} \cdot \nabla B = 0. \qquad\qquad (4.52)$$

## 4.7 Representation of Small-Scale Turbulent Heat Diffusion

The breakdown of the steady linear theory (Section 4.3) indicates that fundamental physical processes are missing from the steady linear model. A process of potential importance is the vertical diffusion of heat by small-scale, three-dimensional turbulent motions, which would introduce a diffusive heat source in the thermodynamic equation (4.7) that could balance the vertical advection of mean temperature by the vertical motion. The planetary geostrophic scaling that results from the a priori specification of the scales of motion (2.1) shows that the direct molecular diffusion of heat on the large-scale fields is negligible. However, the average effect of small-scale (i.e., vertical and horizontal scales of tens of meters or less) turbulent motions on the large-scale field may be many times larger than direct molecular diffusion because the turbulence intensifies the local thermal gradients by orders of magnitude.

A useful, simple representation of the mean effect of small-scale, three-dimensional turbulence on the large-scale motion can be achieved by presuming that the turbulent diffusion acts like molecular diffusion but with an enhanced diffusivity $\kappa_v \gg \kappa_T$. Because the vertical thermal gradients on large scales are typically larger than the horizontal gradients by a factor equal to the aspect ratio $D/L \approx 10^{-3}$, only the vertical divergence of this diffusion need be retained at first order. This yields the modified thermodynamic equation

$$\frac{DT}{Dt} = \kappa_v \frac{\partial^2 T}{\partial z^2}. \qquad\qquad (4.53)$$

Note that if (4.53) is taken to replace (4.10), with $\kappa_v \neq 0$, then the conservation laws (4.49) and (4.52) for potential vorticity $Q$ and Bernoulli function $B$, respectively, no longer hold. The importance of the vertical divergence $\kappa_v \partial^2 T/\partial z^2$ of the vertical turbulent heat flux, relative to the individual terms of the large-scale material advection of heat $DT/Dt$, will depend on the value that is specified for $\kappa_v$.

It is essential to recognize that this addition of a representation of the average effect of small-scale turbulence to the planetary geostrophic model is a fundamental departure from the deductive scaling approach by which the planetary geostrophic equations (2.102)–(2.107) were derived as a relatively rigorous, filtered approximation for the purely large-scale motion. Many other types of small-scale or intermediate-scale motions might equally well have a mean effect on the large-scale flow, which might be introduced into the planetary geostrophic model through other parameterizations. However, because it represents a fundamental process through which heat is forced vertically downward from the warm surface into the cool ocean interior, and because the wide separation of time and space scales between the planetary geostrophic and three-dimensional turbulent motions provides a relatively robust motivation for the simple diffusive model, the particular addition (4.53) merits special attention. This small-scale turbulent mixing is an essential component of the internal boundary layer theory of the main thermocline and the diffusively driven theory of the abyssal circulation (see Chapter 5).

## 4.8 The *M* Equation

With the zonal component of the geostrophic relation (4.7), the Sverdrup vorticity relation (3.8) may be written in the form

$$\frac{1}{\rho_0}\frac{\partial p'}{\partial x} + \frac{\partial}{\partial z}\left(-\frac{f^2}{\beta}w\right) = \operatorname{div}\left(\frac{p'}{\rho_0}, -\frac{f^2}{\beta}w\right) = 0. \tag{4.54}$$

Thus the vector field $(p'/\rho_0, -f^2 w/\beta)$ is divergence free, and a stream function $M(x, z; y, t)$ can be defined so that

$$\frac{p'}{\rho_0} = \frac{\partial M}{\partial z}, \quad \frac{f^2}{\beta}w = \frac{\partial M}{\partial x}. \tag{4.55}$$

From (4.7)–(4.8) and (4.55), it follows that

$$u = -\frac{1}{f}\frac{\partial^2 M}{\partial y \partial z}, \quad v = \frac{1}{f}\frac{\partial^2 M}{\partial x \partial z}, \quad T = \frac{1}{g\alpha_T}\frac{\partial^2 M}{\partial z^2} \tag{4.56}$$

so that the thermodynamic equation (4.53) may be written as

$$\frac{\partial^3 M}{\partial t \partial z^2} - \frac{1}{f}\frac{\partial^2 M}{\partial y \partial z}\frac{\partial^3 M}{\partial x \partial z^2} + \frac{1}{f}\frac{\partial^2 M}{\partial x \partial z}\frac{\partial^3 M}{\partial y \partial z^2} + \frac{\beta}{f^2}\frac{\partial M}{\partial x}\frac{\partial^3 M}{\partial z^3} = \kappa_v\frac{\partial^4 M}{\partial z^4}. \tag{4.57}$$

This shows that the set of planetary geostrophic equations (4.7)–(4.9) with (4.53) are mathematically equivalent to a single nonlinear, fifth-order partial differential equation in the scalar variable $M(x, y, z, t)$.

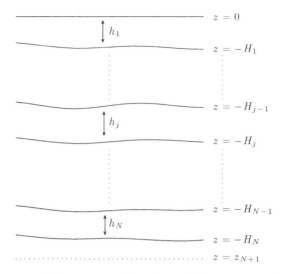

Figure 4.2. Layer-model geometry. The layer thicknesses are $h_j(x, y, t)$, $j = \{1, 2, 3, \ldots, N\}$, the sea surface is at $z = 0$, and the interface between layers $j$ and $j + 1$ is at the depth $z = -H_j(x, y, t)$. The constant level $z = Z_{N+1}$ is in the motionless layer $N + 1$, above the sea floor.

## 4.9 Reduced-Gravity Layer Models

The theory of the adiabatic, advective subtropical thermocline is most conveniently developed in the setting of layered models of the stratified fluid, in which the density, or temperature, stratification is concentrated into infinitesimally thin interfaces between finite-thickness layers that are each of uniform but distinct density or temperature. Consider such a model with $N + 1$ layers: a deep, motionless layer with temperature $T_{N+1}$ and $N$ successively shallower intermediate layers with temperature $T_j$, pressure $p'_j$, thickness $h_j$, and velocity components $u_j, v_j, w_j$, $j = \{N, N - 1, \ldots, 1\}$ (Figure 4.2). The interface between layer $j$ and layer $j + 1$ will be at $z = -H_j$, where $H_j$ is the sum of the layer thickness $h_k$ for $k = \{1, \ldots, j\}$:

$$H_j(x, y, t) = \sum_{k=1}^{j} h_k(x, y, t). \tag{4.58}$$

The temperatures $T_j$, $j = \{1, \ldots, N + 1\}$ are each constant, so the hydrostatic relation (4.8) may be integrated to obtain the pressures $p'_j(z)$, $j = \{1, \ldots, N\}$, $-H_{j+1} < z < -H_j$ in the moving layers:

$$p'_j(z) = p'_{N+1,0} - \rho_0 g \alpha_T (z_{N+1} T_{N+1} - z T_j) + \rho_0 \sum_{k=j}^{N} \gamma_k H_k, \tag{4.59}$$

where $p'_{N+1,0}$ is the constant value of the pressure at $z = z_{N+1} < -H_N$ in the motionless deep layer. The constant

$$\gamma_j = g\alpha_T \left( T_j - T_{j+1} \right) = g\frac{\rho_{j+1} - \rho_j}{\rho_0} \tag{4.60}$$

is the reduced gravity associated with layer $j$, with $j = \{1, \ldots, N\}$ for this $N$-layer reduced-gravity model, and is equal to the gravitational acceleration $g$ reduced by a factor equal to the fractional density difference across the layer interface at the base of the $j$th layer.

The expression (4.59) may be written in terms of the layer-model Bernoulli functions $B_j$, $j = \{1, \ldots, N\}$, where

$$B_j(x, y, t) = \sum_{k=j}^{N} \gamma_k H_k(x, y, t). \tag{4.61}$$

Because, for $z$ in layer $j$, $B(x, y, z, t) - B_j(x, y, t) = p'_{N+1,0}/\rho_0 - g\alpha_T T_{N+1} z_{N+1}$, these layer Bernoulli functions $B_j$ differ from the continuous Bernoulli function $B(x, y, z, t)$ only by an inconsequential constant. Note that unlike the pressure functions $p'_j(x, y, z, t)$, $B_j$ and $H_j$ are functions only of $x$, $y$, and $t$. Because

$$\frac{1}{\rho_0}\nabla_h p'_j = \nabla_h B_j, \quad j = \{1, \ldots, N\}, \tag{4.62}$$

the Bernoulli functions may be used in place of the pressure in the geostrophic relations so that

$$u_j = -\frac{1}{f}\frac{\partial B_j}{\partial y}, \quad v_j = \frac{1}{f}\frac{\partial B_j}{\partial x}, \quad j = \{1, \ldots, N\}. \tag{4.63}$$

The thermal wind equations for the velocity differences between adjacent layers take the form

$$u_j - u_{j+1} = -\frac{\gamma_j}{f}\frac{\partial H_j}{\partial y}, \quad v_j - v_{j+1} = \frac{\gamma_j}{f}\frac{\partial H_j}{\partial x}, \quad j = \{1, \ldots, N\}. \tag{4.64}$$

Because, by (4.63), the horizontal velocity $\mathbf{u}_j = (u_j, v_j)$ is independent of $z$ within each layer, the vertical velocity $w_j(z)$ in layer $j$ will have constant vertical gradient

$$\frac{\partial w_j}{\partial z} = -\nabla_h \cdot \mathbf{u}_j. \tag{4.65}$$

The boundary condition on vertical velocity in layer $j$ at the layer interface $z = -H_j$, $j = \{1, \ldots, N\}$ is, following (3.11) at the base of layer $j$,

$$w_j(z = -H_j) = -\frac{\partial H_j}{\partial t} - \mathbf{u}_j \cdot \nabla_h H_j(x, y, t), \quad j = \{1, \ldots, N\}. \tag{4.66}$$

From (4.64), and with (3.11) at the top of layer $j + 1$, it follows also that

$$w_{j+1}(z = -H_j) = -\frac{\partial H_j}{\partial t} - \mathbf{u}_{j+1} \cdot \nabla_h H_j$$

$$= -\frac{\partial H_j}{\partial t} - \mathbf{u}_j \cdot \nabla_h H_j$$

$$= w_j(z = -H_j), \tag{4.67}$$

so the vertical velocity $w(z)$ is continuous at the layer interfaces. Equation (4.9) for conservation of mass may then be integrated vertically within each layer, yielding

$$\frac{\partial h_1}{\partial t} + \frac{\partial (h_1 u_1)}{\partial x} + \frac{\partial (h_1 v_1)}{\partial y} = -W_E \tag{4.68}$$

$$\frac{\partial h_j}{\partial t} + \frac{\partial (h_j u_j)}{\partial x} + \frac{\partial (h_j v_j)}{\partial y} = 0, \quad j = \{2, \dots, N\}, \tag{4.69}$$

where, by (3.26), the planetary geostrophic vertical velocity $w = W_E$ at $z = 0$.

In each layer, let

$$Q_j = \frac{f}{h_j}, \quad j = \{1, \dots, N\} \tag{4.70}$$

be the potential vorticity, and let

$$\left.\frac{D}{Dt}\right|_j = \frac{\partial}{\partial t} + u_j \frac{\partial}{\partial x} + v_j \frac{\partial}{\partial y}, \quad j = \{1, \dots, N\} \tag{4.71}$$

be the material derivative for layer quantities, such as thickness $h_j$, that are independent of depth $z$ within the layer. Then, for $j \geq 2$,

$$\left.\frac{DQ_j}{Dt}\right|_j = \left.\frac{D}{Dt}\right|_j \frac{f}{h_j}$$

$$= \frac{f}{h_j} \left( \frac{\beta}{f} v_j - \frac{1}{h_j} \left.\frac{Dh_j}{Dt}\right|_j \right)$$

$$= \frac{f}{h_j} \left( \frac{\beta}{f} v_j + \frac{\partial u_j}{\partial x} + \frac{\partial v_j}{\partial y} \right)$$

$$= \frac{f}{h_j} \left( \frac{\beta}{f} v_j - \frac{\partial w_j}{\partial z} \right),$$

so the potential vorticity is conserved following the motion in layer $j$, $j = \{2, \dots, N\}$:

$$\left.\frac{DQ_j}{Dt}\right|_j = 0, \quad j = \{2, \dots, N\}. \tag{4.72}$$

In contrast, if $W_E \neq 0$, then the layer 1 potential vorticity $Q_1$ is not conserved following the layer 1 motion; instead,

$$\left.\frac{DQ_1}{Dt}\right|_2 = \frac{f}{h_1^2} W_E. \tag{4.73}$$

From (4.61)–(4.63) and (4.71), it follows that

$$\left.\frac{DB_j}{Dt}\right|_j = \frac{\partial B_j}{\partial t}, \quad j = \{1, \ldots, N\}, \tag{4.74}$$

so in steady flow, the Bernoulli functions $B_j$, $j = \{1, \ldots, N\}$, are conserved following the motion in the respective layers.

For $j \geq 2$, there is an additional simplification for steady flow: because

$$\frac{\partial(h_j u_j)}{\partial x} + \frac{\partial(h_j v_j)}{\partial y} = 0, \quad j = \{2, \ldots, N\}, \tag{4.75}$$

it follows that a transport stream function $\psi_j$ can be defined in layer $j$ so that

$$h_j u_j = -\frac{\partial \psi_j}{\partial y}, \quad h_j v_j = \frac{\partial \psi_j}{\partial x}, \quad j = \{2, \ldots, N\}. \tag{4.76}$$

Thus, by (4.72) and (4.74), there are functions $\hat{Q}_j$ and $\hat{B}_j$ such that

$$Q_j = \hat{Q}_j(\psi_j), \quad B_j = \hat{B}_j(\psi_j), \quad j = \{2, \ldots, N\}. \tag{4.77}$$

With (4.63) and (4.70), (4.76) implies that $\nabla_h B_j = Q_j \nabla_h \psi_j$, and so also

$$\hat{Q}_j(\psi_j) = \frac{d\hat{B}_j}{d\psi_j}, \quad j = \{2, \ldots, N\}. \tag{4.78}$$

Another useful result for steady flow can be derived from (4.66) and (4.64):

$$\begin{aligned}
w_j(z = -H_j) &= -\mathbf{u}_j \cdot \nabla_h H_j(x, y, t) \\
&= -\mathbf{u}_j \cdot \left[ -\frac{f}{\gamma_j} \mathbf{k} \times (\mathbf{u}_j - \mathbf{u}_{j+1}) \right] \\
&= -\frac{f}{\gamma_j} \mathbf{u}_j \cdot \mathbf{k} \times \mathbf{u}_{j+1} \\
&= -\frac{f}{\gamma_j} \mathbf{k} \cdot \mathbf{u}_{j+1} \times \mathbf{u}_j,
\end{aligned} \tag{4.79}$$

where $\mathbf{k}$ is the vertical unit vector. Thus, in regions such as the subtropical gyre upper thermocline, where the vertical velocity is downward, the horizontal velocity rotates clockwise with depth in the northern hemisphere. In addition, Equation (4.79) allows an estimate of the vertical velocity at the interface $z = -H_j$—which is typically small and difficult or impossible to measure directly—to be obtained from the magnitude and rotation of the geostrophic shear across the interface by integrating the Sverdrup vorticity relation (3.8) vertically from the motionless layer $N + 1$ to $z = -H_j$. This yields the equation

$$\frac{\beta}{f} \sum_{m=j+1}^{N} h_m v_m = w(z = -H_j) = -\frac{f}{\gamma_j} \mathbf{k} \cdot \mathbf{u}_{j+1} \times \mathbf{u}_j. \tag{4.80}$$

The relation (4.80) is known as the $\beta$-spiral. It allows the vertically integrated geostrophic Sverdrup transport between the level of no motion and $z = -H_j$ to be estimated from the magnitude and rotation of the geostrophic velocity at the interface $z = -H_j$.

## 4.10 Notes

Gill (1982) provides the equations and tables for a recent standard seawater equation of state. Linear planetary waves are also known as Rossby waves, and the internal and external deformation radii as Rossby radii, in recognition of the contributions of the meteorologist Carl-Gustav Rossby. The $M$ equation was first derived by Welander (1959). The turbulent diffusion term in (4.53) was first introduced into the planetary geostrophic thermodynamic equation by Robinson and Stommel (1959). Additional general discussion of shallow-water layer models, including detailed descriptions of the associated linear quasi geostrophic and ageostrophic waves, and the role of planetary waves in the spin-up of the wind-forced gyres from a state of rest, is given by Pedlosky (1987) and Gill (1982).

# 5

# Circulation in a Simple Rectangular Basin

## 5.1 Domain and Boundary Conditions

With the basic large-scale-approximate equations and some of their general properties established, it is appropriate now to examine the circulation and thermocline structures that arise in specific solutions of these equations. Even for the simplest case of steady solutions, in which the flow is independent of time, this requires that a sufficient set of boundary conditions be specified. This set of boundary conditions is not unique: a variety of such conditions, which differ in physical and mathematical detail, may yield solutions of physical interest for the large-scale flow. In general, these sets of conditions for idealized models of ocean gyre structure and circulation must be chosen to represent the definitive characteristics of the general physical setting in which these large-scale features develop.

A fundamental element of the boundary conditions is the geometry specified for the ocean basin. The simplest such basin is a rectangular domain, restricted to a portion of one hemisphere, with vertical sidewalls aligned along lines of constant longitude $x = \{x_W, x_E\}$ and latitude $y = \{y_S, y_N\}$ and a flat bottom at the constant depth $z = -H_0$ (Figure 5.1). Such a choice avoids the singularity of the planetary geostrophic momentum equations that arises at the equator, where the Coriolis parameter $f$ vanishes. It also removes a complicating geometric element: the circumpolar connection that exists around the Antarctic continent in the southern hemisphere, which will be seen (Chapters 7 and 8) to have significant impact on the surface and mid-depth circulation. The use of vertical rather than sloping sidewalls may be motivated on the grounds that the lateral scale of the continental slope regions is generally small compared to the scale of ocean basins.

In this geometry, the simplest surface thermal conditions—which, for these reduced, temperature-only equations, effectively include the influence of salinity or freshwater fluxes as well as heat fluxes—must represent the basic meridional gradient in surface heating, with warmer temperatures and less dense surface waters at subtropical

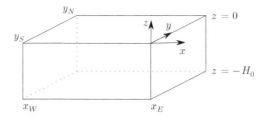

Figure 5.1. Domain geometry for the rectangular, closed, single-hemisphere basin. The lateral boundaries are vertical, with rigid walls at $x = x_E$, $x = x_W$, $y = y_S$, and $y = y_N$. The rigid, flat top and bottom are at $z = 0$ and $z = -H_0$, respectively.

latitudes and cooler temperatures and denser water at subpolar latitudes. Similarly, the wind stress forcing field at the sea surface must represent the equatorial and tropical easterlies, and the midlatitude westerlies.

At the rigid boundaries of the rectangular domain, the standard no-normal-flow condition $\mathbf{u} \cdot \mathbf{n}$ will generally apply, where $\mathbf{n}$ is the unit normal to the boundary, requiring that

$$u = 0 \quad \text{at } x = \{x_E, x_W\}, \quad v = 0 \quad \text{at } y = \{y_S, y_N\} \tag{5.1}$$

$$w = 0 \quad \text{at } z = -H_0. \tag{5.2}$$

As discussed earlier in connection with the depth-integrated flow (Section 3.7), however, even this simple set of conditions cannot be satisfied in general by the planetary geostrophic equations, which must be supplemented by additional terms that support, at least, an appropriate western boundary layer. If no-slip conditions are to be imposed instead, for which the tangential as well as normal components of velocity vanish, then boundary layers are generally required at all rigid boundaries. Most of the available analytical theory for the stratified planetary geostrophic equations, including that discussed here, does not include explicit treatment of the western boundary layer and thus is formally incomplete. General assumptions about the behavior of the flow along the western boundary are frequently made, based on physical intuition and the character of the boundary layers that arise for simple friction schemes in the depth-integrated flow. In some cases, the implications of these assumptions can be explored quantitatively.

For most of the solutions considered here, the surface wind stress forcing will be imposed through the Ekman transport balance and the proportionality of vertical velocity to the curl of the wind stress (3.26), while the surface temperature distribution will be specified directly:

$$T(x, y, 0, t) = T_a(x, y, t). \tag{5.3}$$

The specified temperature field $T_a(x, y, t)$ may be considered to represent a surface air temperature field with which the temperature of the ocean surface is locally in equilibrium. More generally, a surface heat flux $Q_h$ may be imposed that is proportional to the difference between the ocean surface temperature $T_s(x, y, t) = T(x, y, 0, t)$ and the imposed surface air temperature $T_a$,

$$Q_h(x, y) = -\gamma_a(T_s - T_a), \tag{5.4}$$

where $\gamma_a$ is an inverse time scale. The imposed ocean surface temperature condition (5.3) may be interpreted as the limit $\gamma_a \to \infty$ of (5.4), in which the relaxation time scale is small compared to the time scale of the flow.

As a specific example of idealized forcing functions, $W_E(x, y)$ and $T_a(y)$ may be taken as

$$W_E(y) = W_0 \cos\left[\frac{2\pi(y - y_S)}{y_N - y_S}\right], \quad T_a(y) = T_S + \frac{T_N - T_S}{y_N - y_S}(y - y_S), \tag{5.5}$$

where $W_0 > 0$, $T_S$, and $T_N$ are suitable constants. In this case, there is a subtropical gyre regime with $W_E < 0$ in the region $(y_N - y_S)/4 < y < 3(y_N - y_S)/4$ and a subpolar gyre regime with $W_E > 0$ in the region $3(y_N - y_S)/4 < y < y_N$. There is also a tropical upwelling regime with $W_E > 0$ for $y_S < y < (y_N - y_S)/4$, which may be taken to represent loosely the equatorial Ekman transport divergence and upwelling that results from the change in sign of $f$ at the equator in the presence of mean easterly wind stress. An equivalent zonal wind stress structure can be associated with this form of $W_E(y)$, according to

$$\tau_w^x(y) = -\rho_0 f \int_{y_S}^{y} W_E(y')dy', \tag{5.6}$$

giving the familiar pattern of low-latitude easterlies and midlatitude westerlies. Note that $\partial \tau_w^x / \partial y \neq 0$ at $y = y_S$ and $y = y_N$; for the depth-integrated flow with linear friction (3.40), there are then meridional boundary layers of width $[(x_E - x)r/\beta]^{1/2}$ near $y = y_S$ and $y = y_N$. With $T_N < T_S$, the coldest and densest fluid will be produced by surface cooling in the subpolar gyre, near $y = y_N$. If this surface cooling results in density inversions, with denser fluid overlying less dense fluid, convective instabilities would in general be excited, but such motions have largely been eliminated from direct representation in the planetary geostrophic approximation. Accordingly, the effects of these convective motions must generally be represented by ad hoc vertical rearrangements of the fluid column—convective adjustment—rather than by explicit solutions of the planetary geostrophic equations.

## 5.2 Planetary Geostrophic Energetics

Before considering explicit solutions, it is useful to consider the energetics of planetary geostrophic motion. An evolution equation for the potential energy $\mathcal{P}$ may be constructed by taking the time derivative of the integral of the potential energy density

$g\rho z$ over the volume $V_0$ of the ocean basin, where $\rho = -\rho_0 \alpha_T T$:

$$\frac{d\mathcal{P}}{dt} = \frac{d}{dt} \left( \int_{V_0} g\rho z \, dV \right)$$

$$= \frac{d}{dt} \left( -\int_{V_0} g\rho_0 \alpha_T T z \, dV \right)$$

$$= -g\rho_0 \alpha_T \left( \int_{V_0} T \frac{Dz}{Dt} \, dV + \int_{V_0} \frac{DT}{Dt} z \, dV \right)$$

$$= -g\rho_0 \alpha_T \left[ \int_{V_0} wT \, dV - \kappa_v \int_{A_0} \left( T|_{z=0} - T|_{z=-H_0} \right) dx \, dy \right]. \quad (5.7)$$

Here turbulent diffusion of heat has been included in the thermodynamic balance (4.53), the turbulent flux of heat through the ocean bottom is taken to be zero, and the volume element $dV = dx \, dy \, dz$.

A kinetic energy equation may be obtained by multiplying each of the momentum equations (4.7)–(4.8) by the corresponding velocity and integrating over $V_0$:

$$0 = \int_{V_0} \left[ u \left( \rho_0 f v - \frac{\partial p'}{\partial x} \right) + v \left( \rho_0 f u - \frac{\partial p'}{\partial y} \right) + w \left( -\frac{\partial p'}{\partial z} + g\rho_0 \alpha_T T \right) \right] dV$$

$$= \int_{V_0} \left( -u \frac{\partial p'}{\partial x} - v \frac{\partial p'}{\partial y} - w \frac{\partial p'}{\partial z} + g\rho_0 \alpha_T w T \right) dV$$

$$= \int_{V_0} \left[ -\nabla \cdot (\mathbf{u} p') + g\rho_0 \alpha_T w T \right] dV$$

$$= -\int_{A_0} W_E \, p' \, dx \, dy - \int_{A_0} w_* p' \, dx \, dy + g\rho_0 \alpha_T \int_{V_0} w T \, dV. \quad (5.8)$$

Here, because $\mathbf{u}$ represents the planetary geostrophic flow alone, the upper boundary condition from (3.25) and (3.27) is $w(z = 0) = W_E + w_*$, where $W_E$ is the surface Ekman pumping velocity and $w_*$ is the difference of evaporation and precipitation rates. The no-normal-flow condition $\mathbf{u} \cdot \mathbf{n} = 0$ has been assumed at all solid boundaries, and $A_0$ denotes the surface area of the ocean basin at $z = 0$. Because

$$W_E = \nabla_h \cdot \mathbf{U}_E = \frac{1}{\rho_0} \left[ \frac{\partial}{\partial x} \left( \frac{\tau_w^y}{f} \right) - \frac{\partial}{\partial y} \left( \frac{\tau_w^x}{f} \right) \right], \quad (5.9)$$

the first integral in (5.8) may be written as

$$-\int_{A_0} W_E \, p' \, dx \, dy = -\int_{A_0} \nabla_h \cdot (p' \mathbf{U}_E) \, dx \, dy + \int_{A_0} \mathbf{u}_h \cdot \boldsymbol{\tau}_w \, dx \, dy, \quad (5.10)$$

where $(U_E, V_E)$ are the Ekman transports (3.21) and where (3.9) has been used. If there is no Ekman transport through the lateral boundaries—which, for these large-scale momentum equations, requires that the tangential wind stress along the boundaries vanish—the first of these two integrals vanishes. In that case, the kinetic energy equation (5.8) reduces to three terms: the area-integrated product of the surface velocity and surface stress; the area-integrated product of the precipitation-evaporation difference and surface pressure; and the volume-integrated vertical flux of buoyancy:

$$\int_{A_0} \mathbf{u}_h \cdot \boldsymbol{\tau}_w \, dx \, dy - \int_{A_0} w_* p' \, dx \, dy + g\rho_0\alpha_T \int_{V_0} wT \, dV = 0. \qquad (5.11)$$

The latter term, proportional to $\int wT dV$, appears also in (5.7) and represents the conversion between potential and kinetic energy, as relatively denser or less dense waters are moved vertically, altering the mass field and thus the volume-integrated potential energy. Note that the balance (5.11) is diagnostic: there is no kinetic energy storage term because the large-scale approximations imply that the kinetic energy is negligible relative to the potential energy.

The total energy equation is obtained as the sum of (5.7) and (5.11):

$$\frac{d\mathcal{P}}{dt} = \int_{A_0} \mathbf{u}_h \cdot \boldsymbol{\tau}_w \, dx \, dy - \int_{A_0} w_* p' \, dx \, dy$$

$$+ g\rho_0\alpha_T\kappa_v \int_{A_0} \left( T|_{z=0} - T|_{z=-H_0} \right) dx \, dy. \qquad (5.12)$$

The first term on the right-hand side of (5.12) may, in principle, be of either sign; in general, however, the large-scale surface velocities in the subtropical gyres and the Antarctic Circumpolar Current are roughly aligned with the great bands of tropical easterlies and midlatitude westerlies so that this integral over the world ocean is positive, providing the energy source for the wind-driven circulation. As might be inferred from large-scale surface salinity distribution (Figure 1.2), evaporation typically dominates in the high-pressure subtropical latitudes, and precipitation dominates in the low-pressure high latitudes, so the second term on the right-hand side is typically negative. However, it is also relatively small because of the small effective precipitation and evaporation velocities. Finally, because surface temperatures are generally greater than bottom temperatures, the last term is also positive; it is through this term that the turbulent diffusion represented by $\kappa_v$ becomes a source of energy for the diffusively driven circulation.

Thus the dominant terms in (5.12) are source terms, and the energetics do not balance. In numerical solutions in closed basins, energetic balance is generally achieved through the same additional friction and diffusion terms that are added to the momentum and thermodynamic equations to support the needed lateral boundary layers and through convective adjustment schemes. The convective adjustment terms are especially critical to the global heat and energy balance. They must be introduced because the planetary geostrophic dynamics allow density inversions, in which denser fluid overlies less dense fluid, to be maintained on long time scales, a situation that does not persist in nature because of convective or gravitational instabilities that have been largely filtered out of the planetary geostrophic dynamics.

Convective adjustment schemes modify the vertical profile of density in such cases by rearranging or mixing the fluid column so that the stratification is neutral or positive. For the thermocline equations (4.7)–(4.10), this means that at each $(x, y)$, the temperature profile $T(x, y, z)$ must satisfy

$$\frac{\partial T}{\partial z} \geq 0 \quad \text{for } -H_0 \leq z \leq 0. \tag{5.13}$$

The vertical fluxes implied by the associated instantaneous vertical adjustment are generally required to preserve the total heat content,

$$\int_{-H_0}^{0} T_{\text{after}} \, dz = \int_{-H_0}^{0} T_{\text{before}} \, dz, \tag{5.14}$$

where $T_{\text{after}}$ and $T_{\text{before}}$ are the temperature profiles before and after the adjustment, respectively. Because the effect of this rearrangement is always to raise less dense (warmer) fluid and denser (cooler) fluid, it follows that

$$\int_{-H_0}^{0} z \, T_{\text{after}} \, dz \geq \int_{-H_0}^{0} z \, T_{\text{before}} \, dz. \tag{5.15}$$

Thus such schemes always decrease the potential energy of the large-scale state. Because ocean convection occurs on relatively small horizontal scales, it is generally assumed that there is no resulting source of mechanical energy for the large-scale flow, and it is only through the rearrangement of the mass field that the convective adjustment parameterization is allowed to act. The potential energy that is thereby released from the large-scale field is presumed to drive the small-scale convective motion that would accomplish the parameterized vertical adjustment or to be dissipated locally. Convective adjustment schemes of this general type may be included as a component of a numerical solution procedure for the large-scale equations, or the condition (5.13) may be invoked directly as a constraint on analytical calculations. For a general equation of state, the stability condition (5.13) must be replaced by $N^2 \geq 0$, where $N^2$ is given by (2.74).

## 5.3 Thermocline Scaling

The warm water of the upper and main subtropical thermoclines extends from the tropics to midlatitudes and from the surface to a depth of roughly 500–800 m in all the subtropical ocean basins (Figure 1.4). The corresponding band of latitudes is roughly demarcated by the easterly trade winds on the equatorward sides and by the midlatitude westerly winds on the poleward sides. This layer of warm water is relatively shallow, covering at its deepest points no more than one-fifth of the ocean's mean depth. Thus it has the character of a surface-trapped boundary layer.

Scaling estimates can be used to obtain the corresponding characteristic depth scale for this boundary layer and provide an estimate of thermocline depth and its dependence on parameters. In general, one might hope to use an associated boundary layer balance to simplify further the equations (4.7)–(4.9) with (4.53) or the equivalent, single, fifth-order partial differential equation (4.57). However, with the exception of the choice regarding inclusion or neglect of the turbulent heat diffusion in (4.53), no additional simplifications are possible, and the scales instead reflect a balance of all the included terms; that is, the planetary geostrophic equations themselves may already be considered to form a set of simplified boundary layer equations for the subtropical main thermocline. Nonetheless, it is still possible to derive useful thermocline depth scales by this approach.

First, consider the adiabatic large-scale asymptotic equations (4.7)–(4.10), without the turbulent heat diffusivity $\kappa_v$. Let $D_a$ be the unknown scale depth of the thermocline, and let $V_a$ and $P_a$ be the corresponding unknown horizontal geostrophic velocity and hydrostatic pressure scales, respectively. Let the vertical velocity scale be set by the Ekman pumping velocity from (3.26), for which a characteristic value is $W_E = W = 10^{-6}$ m s$^{-1}$ $\approx 30$ m yr$^{-1}$, as in (2.1). Then, with other scales also as in (2.1), Equations (4.7)–(4.10) may be replaced by scaling balances:

$$f_0 V_a = \frac{P_a}{\rho_0 L}, \tag{5.16}$$

$$\frac{P_a}{D_a} = \rho_0 g \alpha_T \Delta T_s, \tag{5.17}$$

$$\frac{V_a}{L} = \frac{W_E}{D_a}, \tag{5.18}$$

$$\frac{V_a \Delta T_s}{L} = \frac{W_E \Delta T_s}{D_a}, \tag{5.19}$$

the last two of which are redundant. Here the surface temperature difference $\Delta T_s$ has been used for the characteristic temperature scale under the assumption that the temperature of the deep fluid beneath the thermocline is approximately equal to that of the coldest surface fluid. Note that the continuity equation scaling (5.18) can be replaced by a scaling of the Sverdrup vorticity relation (3.2) without changing the essential result since $f_0 \approx \beta L$.

Equations (5.16)–(5.19) may be solved to obtain

$$D_a = \left( \frac{f_0 L^2 W_E}{g \alpha_T \Delta T_s} \right)^{1/2}, \quad V_a = \frac{W_E L}{D_a}, \quad P_a = g \alpha_T \Delta T_s D_a. \tag{5.20}$$

Because the thermodynamics are purely adiabatic and advective, the depth scale $D_a$ is called the advective scale for thermocline depth. Through the Ekman pumping velocity scale $W_E$, it is proportional to the square root of the amplitude of the wind stress. Substituting characteristic dimensional values for the parameters in (5.20) yields the dimensional scales

$$D_a \approx 500 \text{ m}, \quad V_a \approx 10^{-2} \text{ m s}^{-1}. \tag{5.21}$$

Thus the advective scale $D_a$ for thermocline depth is comparable to the observed depth of the main thermocline in the subtropical gyres (Figures 1.4–1.7).

Alternatively, an advective-diffusive scale for thermocline depth can be obtained by including the turbulent diffusion term, proportional to $\kappa_v$, in the thermodynamic equation, as in (4.53), and allowing the corresponding vertical velocity scale $W_\delta$ to be determined from the scaled balances rather than imposed. Let $\delta$ be the unknown advective-diffusive scale depth of the thermocline, and let $V_\delta$ and $P_\delta$ be the corresponding unknown horizontal geostrophic velocity and hydrostatic pressure scales, respectively. Then the scaling balances are

$$f_0 V_\delta = \frac{P_\delta}{\rho_0 L}, \tag{5.22}$$

$$\frac{P_\delta}{D_\delta} = \rho_0 g \alpha_T \Delta T_s, \tag{5.23}$$

$$\frac{V_\delta}{L} = \frac{W_\delta}{D_\delta}, \tag{5.24}$$

$$\left[ \frac{V_\delta \Delta T_s}{L}, \frac{W_\delta \Delta T_s}{D_a} \right] = \kappa_v \frac{\delta T_s}{\delta^2}. \tag{5.25}$$

Equations (5.22)–(5.25) may be solved to obtain

$$\delta = \left( \frac{f_0 L^2 \kappa_v}{g \alpha \Delta T_s} \right)^{1/3}, \quad V_\delta = \frac{\kappa_v L}{\delta^2}, \quad W_\delta = \frac{\kappa_v}{\delta}, \quad P_\delta = g \alpha_T \Delta T_s \delta. \tag{5.26}$$

The advective-diffusive scale $\delta$ is thus proportional to $\kappa_v^{1/3}$, and the induced vertical velocity $W_\delta$ is proportional to $\kappa_v^{2/3}$. A suitable estimate of $\kappa_v$ in the subtropical thermocline, based on measurements of turbulent dissipation and tracer dispersion, is $\kappa_v = 1.5 \times 10^{-5} \text{ m}^2 \text{ s}^{-1} \approx 100 \kappa_T$, which gives

$$\delta \approx 150 \text{ m}, \quad V_\delta \approx 0.3 \times 10^{-2} \text{ m s}^{-1}, \quad W_\delta \approx 10^{-7} \text{ m s}^{-1}. \tag{5.27}$$

Thus, with this estimate of $\kappa_v$, the advective-diffusive scale gives a scale depth that is several times smaller than the observed depth of the main thermocline in the subtropical gyres. In addition, $W_\delta$ is an order of magnitude smaller than $W_E$.

Because $W_\delta \ll W_E$, and $\delta$ is several times smaller than $D_a$, it may be anticipated that the primary effect of the turbulent diffusion $\kappa_v$ is felt only in an internal boundary layer at the base of the wind-driven layer, where the vertical velocity associated with the wind-driven geostrophic motion vanishes. In this case, the horizontal velocity and the horizontal slope of the isotherms will be fixed by the advective, wind-driven motion. With this modification, the diffusive scaling equations (5.22)–(5.25) become

$$f_0 V_a / \delta_i = \frac{g \alpha_T \Delta T_s}{l_i}, \tag{5.28}$$

$$\frac{V_a}{L} = \frac{W_i}{\delta_i}, \tag{5.29}$$

$$\left[ \frac{V_a \Delta T_s}{L}, \frac{W_i \Delta T_s}{\delta_i} \right] = \kappa_v \frac{\Delta T_s}{\delta_i^2}, \tag{5.30}$$

where $\delta_i$ is the unknown internal boundary layer thickness scale and $V_i$ and $W_i$ are the corresponding horizontal and vertical velocity scales, respectively. In the thermal wind equation (5.28), which replaces the geostrophic and hydrostatic relations, the horizontal scale $l_i$ is fixed by the vertical scale $\delta_i$ and the isotherm slope $D_a/L$ from the wind-driven motion so that $l_i = \delta_i (D_a/L)$. These equations may be solved to obtain

$$\delta_i = \left( \frac{f_0 L^2 \kappa_v}{g \alpha \Delta T_s D_a} \right)^{1/2} = \left( \frac{\kappa_v D_a}{W_E} \right)^{1/2}, \quad W_i = \frac{\kappa_v}{\delta_i}. \tag{5.31}$$

Thus the internal boundary layer thermocline thickness scale $\delta_i$ and vertical velocity scale $W_i$ are both proportional to $\kappa_v^{1/2}$. The estimate $\kappa_v = 1.5 \times 10^{-5}$ m$^2$ s$^{-1}$ gives

$$\delta_i \approx 90 \text{ m}, \quad W_i \approx 2 \times 10^{-7} \text{ m s}^{-1}. \tag{5.32}$$

In this case, $\delta_i$ and $W_i$ are both five times smaller than the respective advective scales $D_a$ and $W_E$. This is consistent with the inferences that advective processes dominate near the surface, and that the wind-driven Sverdrup balance sets the horizontal velocity scale, while the smaller diffusive circulation becomes significant only near the base of the wind-driven motion. The separation of scales is less than an order of magnitude, however, so the associated circulations will not be entirely independent. Nonetheless, the separation is sufficient to motivate the consideration first of a purely adiabatic, wind-driven theory and then the subsequent introduction of diffusive effects as a small perturbation to the adiabatic circulation.

In review of these scaling results, it may be noted that the scales $V_a$ and $D_a$ differ by factors of 10 from the corresponding scales $U$ and $D$ in Section 2.1 that were used in the derivation of the planetary geostrophic approximation. Use of these modified values in that derivation gives the same result so that the approach remains internally consistent. The same is true for the sets of scales arising in the advective-diffusive and internal boundary layer scalings, provided that the full corresponding set of modified depth and velocity scales is used in each case.

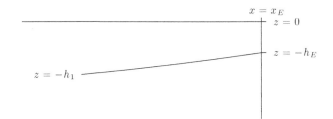

Figure 5.2. Schematic zonal cross section vs. longitude $x$ and depth $z$ for the one-layer reduced-gravity model with layer thickness $h_1$ in the subtropical gyre. At the eastern boundary $x = x_E$, the layer thickness is $h_1(x_E, y) = h_E$ and the base of the layer is at $z = -h_E$.

## 5.4 The One-Layer Model

The warm waters of the subtropical upper thermocline may be most simply represented as a single, homogeneous layer of warm fluid in steady motion, subject to the action of a time-independent wind stress at the surface and lying above a deep motionless layer of cold fluid (Figure 5.2). This is a reduced-gravity layer model with one active layer: $N = 1$. For simplicity, let the eastern and western boundaries be lines of constant longitude, at $x = x_E$ and $x = x_W$, respectively.

For $N = 1$, $H_1 = h_1$, and the horizontal momentum equations are

$$u_1 = -\frac{1}{f}\frac{\partial B_1}{\partial y} = -\frac{\gamma_1}{f}\frac{\partial h_1}{\partial y}, \quad v_1 = \frac{1}{f}\frac{\partial B_1}{\partial x} = \frac{\gamma_1}{f}\frac{\partial h_1}{\partial x}. \tag{5.33}$$

Thus the vertically integrated geostrophic meridional transport is

$$h_1 v_1 = \frac{\gamma_1}{2f}\frac{\partial h_1^2}{\partial x}, \tag{5.34}$$

and the geostrophic Sverdrup transport relation (3.28) may be written as

$$\frac{\beta \gamma_1}{2f^2}\frac{\partial h_1^2}{\partial x} = w|_{z=0} = W_E. \tag{5.35}$$

This may be integrated westward from the eastern boundary $x = x_E$ to obtain the solution for $h_1$,

$$h_1(x, y) = \left[h_E^2 + D_1^2(x, y)\right]^{1/2}, \tag{5.36}$$

where the eastern boundary depth $h_E$ of the warm layer is a boundary condition that must be provided, and the function

$$D_j^2(x, y) = \frac{2f^2}{\beta \gamma_j}\int_{x_E}^{x} W_E(x', y)\, dx'$$

$$= \frac{2f^2}{\beta \gamma_j \rho_0}\int_{x_E}^{x}\left[\frac{\partial}{\partial x}\left(\frac{\tau_w^y}{f}\right) - \frac{\partial}{\partial y}\left(\frac{\tau_w^x}{f}\right)\right] dx' \tag{5.37}$$

can be computed directly from the imposed wind forcing, with $j = 1$ in (5.37) for the one-layer model (5.36). Note that by (5.33), the no-normal-flow condition $u = 0$ at the eastern boundary $x = x_E$ implies that $\partial h_1(x_E, y)/\partial y = 0$, so it is consistent to take $h_E = $ constant.

The solution (5.36)–(5.37) shows that in the subtropical gyre, where the Ekman transport is convergent and the Ekman pumping is downward ($W_E < 0$), the depth $h_1$ of the warm layer increases westward from the eastern boundary. The depth of the warm layer also increases with meridional distance from the equatorward and poleward edges of the gyre, where $W_E = 0$, toward the center of the gyre, where the magnitude $W_E$ of the downward Ekman pumping is maximum. Thus the depth $h_1$ has a half-bowl shape, reaching its maximum depth adjacent to the western boundary in the meridional center of the gyre. Because $f_0 \approx \beta L$, dimensional values of $D_1$ will be comparable to those of $D_a$ in (5.20), with the same square root dependence on $W_E$.

The solution (5.36) does not satisfy a no-normal-flow condition at the western boundary because in general, $\partial h_1(x_W, y)/\partial y \neq 0$. For this one-layer model, a simple frictional boundary layer similar to that in (3.52) could be included; however, analytical extensions of even this simple frictional closure quickly become difficult in multilayer models. Instead, it is assumed here that western boundary layers can exist that close the circulation. It should be recognized that any influence such boundary layers may have on the interior solution will likely not be properly represented in the resulting circulation theories.

At the subpolar boundary $y = y_b$ of the subtropical gyre, the Ekman pumping $W_E$ vanishes, by definition, so that $h_1 = h_E$, and there is no interior geostrophic flow. If the wind stress is purely zonal, however, the equatorward Ekman transport will have a local maximum at $y = y_b$. Thus there is ageostrophic transport across the gyre boundary into the subtropical gyre. If the meridional flow in the western boundary layer is geostrophic, then the total zonally integrated geostrophic meridional transport is

$$M_G = \int_{x_W}^{x_E} h_1 v_1 \, dx = \frac{\gamma_1}{2f} \left( h_E^2 - h_W^2 \right). \tag{5.38}$$

Here $h_W = h(x_W, y_b)$ is the depth at $y = y_b$ of the layer at the western boundary, west of the western boundary layer (Figure 5.3). In steady state, for the closed basin, this poleward geostrophic transport must balance the equatorward Ekman transport across the gyre boundary,

$$M_E = \int_{x_W}^{x_E} V_E \, dx = -\int_{x_W}^{x_E} \frac{\tau_w^x}{\rho_0 f} \, dx, \tag{5.39}$$

so that the volume of warm, subtropical gyre fluid remains constant. If the minimum values of $h_W$ and $h_E$ are chosen that satisfy $M_G + M_E = 0$, the result is

$$h_W = 0, \quad h_E = \left( \frac{2}{\gamma_1 \rho_0} \int_{x_W}^{x_E} \tau_w^x \, dx \right)^{1/2}. \tag{5.40}$$

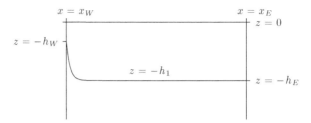

Figure 5.3. Schematic zonal cross section vs. longitude $x$ and depth $z$ for the one-layer reduced-gravity model at the subtropical-subpolar gyre boundary $y = y_b$. At the eastern boundary $x = x_E$ and in the interior $x_W < x < x_E$, the layer thickness is $h_1(x_E, y_b) = h_E$ and the base of the layer is at $z = -h_E$. At the western boundary $x = x_W$, the layer thickness is $h_1(x_W, y_b) = h_W$ and the base of the layer is at $z = -h_W$. The transition from the interior thickness $h_E$ to the western boundary thickness $h_W$ occurs in a narrow western boundary layer, which supports geostrophic flow across the gyre boundary.

This closes the problem for $h_1$, completing the solution (5.36) in terms of the imposed forcing and specific parameters. Note that with this closure, the scale estimate for $h_E$ differs from those for $D_a$ and $D_1$ only by the factor $(L_\tau/L)^{1/2} \approx 1/2$ and thus has a similar square root dependence on the amplitude of the wind forcing.

The choice of the minimum $h_W$ and $h_E$ in the closure (5.40) can be motivated by appeal to a thermodynamic argument: in the absence of forcing, the warm water in the subtropical thermocline would tend to drain to the polar regions, where it would cool, and thus the existing pool of warm water should be the minimum volume that is maintained by the forcing. On the other hand, other balances are certainly possible; for example, this closure neglects a similar balance of Ekman and geostrophic transports that must evidently be operating at the equatorward boundary of the subtropical gyre. Its value lies at least in part in its illustration of the role that geostrophic intergyre transports can play as constraints on basic elements of gyre and thermocline structure.

## 5.5 Multilayer Models: The Ventilated Thermocline

### *A Two-Layer Model*

The temperature of the surface waters of the subtropical gyres generally increases equatorward, while the surface density increases poleward (Figures 1.1 and 1.3). The lateral temperature and density variations associated with these meridional surface gradients are similar in magnitude to the vertical temperature and density variations across the subtropical upper main thermocline. Thus the representation of the warm water volume of the subtropical gyres by a single fluid layer of homogeneous temperature and density is ultimately inadequate even as an idealized model of the qualitative structure of the subtropical gyre circulation.

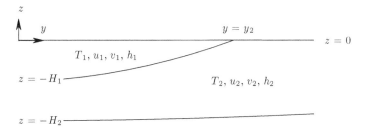

Figure 5.4. Schematic meridional cross section vs. latitude $y$ and depth $z$ for the two-layer ventilated thermocline model of the upper subtropical gyre. North of the outcrop latitude $y = y_2$, the layer 1 thickness $h_1$ vanishes, and layer 2 is the only active layer. South of $y = y_2$, $h_1 > 0$, and there are two active layers: layer 1, which is locally exposed to surface forcing, and layer 2, which is locally shielded from surface forcing by layer 1.

The simplest model that represents the lateral as well as vertical structure of the subtropical thermocline is a two-layer model, in which the warmer of the two moving layers is confined equatorward of a fixed latitude, poleward of which only the colder layer exists. This dividing latitude is the most equatorward latitude at which the colder layer outcrops at the surface. The lateral temperature and density gradients at the surface are represented in the model by the discontinuous surface temperature and density changes across the layer interface at this outcrop latitude.

For the two-layer model, $N = 2$ in (4.59), and the deep motionless layer is layer $N + 1 = 3$. Let $y = y_2$ be the outcrop latitude so that for $y > y_2$ (in the case of a northern hemisphere gyre), the layer 1 thickness $h_1$ vanishes, and layer 2 is in contact with the surface, while for $y < y_2$, there are two moving layers, with layer 1 in contact with the surface and layer 2 insulated below (Figure 5.4).

For $y > y_2$, where there is only a single layer in motion, the solution for the moving-layer thickness $h_2(x, y)$ can be obtained from the single-layer solution (5.36) by replacing the index $j = 1$ everywhere with the index $j = 2$ for layer 2. Thus, in this model, any changes in gyre structure because of the appearance of the second, warmer surface layer at $y = y_2$ are confined to $y < y_2$, equatorward of the outcrop. Note that if, for example, the western boundary layer were allowed to influence the interior, the northward flow of warm water in the western boundary current could alter the interior gyre structure for $y > y_2$.

For $y < y_2$, there are now two layers in motion. The depth-integrated geostrophic transport is

$$h_1 v_1 + h_2 v_2 = \frac{1}{f} \left[ h_1 \left( \gamma_2 \frac{\partial H_2}{\partial x} + \gamma_1 \frac{\partial h_1}{\partial x} \right) + h_2 \gamma_2 \frac{\partial H_2}{\partial x} \right]$$
$$= \frac{\gamma_2}{2f} \frac{\partial}{\partial x} \left( H_2^2 + \Gamma_1 H_1^2 \right), \qquad (5.41)$$

where, in general for $j = 1, \ldots N$,

$$\Gamma_j = \frac{\gamma_j}{\gamma_{j+1}}, \tag{5.42}$$

and the geostrophic Sverdrup transport relation (3.30) may be written as

$$\frac{\beta \gamma_2}{2 f^2} \frac{\partial}{\partial x} \left( H_2^2 + \Gamma_1 H_1^2 \right) = W_E. \tag{5.43}$$

The no-normal-flow condition on the eastern boundary $x = x_E$ implies that

$$H_1(x_E, y) = 0, \quad H_2(x_E, y) = h_E \tag{5.44}$$

because $h_1 = H_1 = 0$ for $y > y_2$, so the Sverdrup transport relation may be integrated to obtain

$$H_2^2 + \Gamma_1 H_1^2 = h_E^2 + D_2^2(x, y). \tag{5.45}$$

Here, for the two-layer model, $D_2^2(x, y)$ is computed from (5.37) with $j = 2$. This does not complete the solution, however, as a second relation is needed to solve separately for $H_1$ and $H_2$ or, equivalently, to determine how the geostrophic Sverdrup transport is distributed vertically over layers 1 and 2.

### The Ventilated Regime

In the region $y < y_2$, equatorward of the outcrop, layer 2 is shielded from the action of the wind by layer 1 above it. In the central part of this region, the required relation between $H_1$ and $H_2$ can be obtained from the conservation properties of the flow in layer 2.

For steady flow, the potential vorticity and Bernoulli equations (4.72) and (4.74) reduce to

$$\mathbf{u}_2 \cdot \nabla_h Q_2 = \mathbf{u}_2 \cdot \nabla_h H_2 = 0 \tag{5.46}$$

which, with the stream function $\psi_2$ defined in (4.76), implies

$$Q_2 = \hat{Q}_2(\psi_2), \quad B_2 = \hat{B}_2(\psi_2). \tag{5.47}$$

Let $(x, y)$ be a point equatorward of the outcrop so that $y < y_2$, and suppose that a layer 2 trajectory can be followed backward (poleward) from $(x, y)$ to a point $(x', y_2)$ at the outcrop, where $x_W < x' < x_E$ (Figure 5.5). By (5.46), the Bernoulli function is conserved along this trajectory so that $H_2(x, y) = H_2(x', y_2)$. At $y = y_2$, $h_1 = 0$, and thus

$$H_2(x, y) = H_2(x', y_2) = h_2(x', y_2). \tag{5.48}$$

At $(x', y_2)$, the potential vorticity $Q_2(x', y_2) = f_2 / h_2(x', y_2)$, where $f_2 = f(y_2)$ is the value of the Coriolis parameter $f$ at the latitude $y = y_2$. By (5.46), $Q_2$ is also

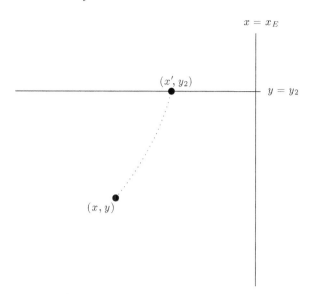

Figure 5.5. Schematic fluid-parcel trajectory in the subsurface layer 2, connecting the unknown point $(x', y_2)$ at the outcrop latitude $y = y_2$, where the layer 1 thickness vanishes, to the given point $(x, y)$.

conserved along the trajectory so that

$$Q_2(x, y) = \frac{f(x, y)}{h_2(x, y)} = Q_2(x', y_2) = \frac{f_2}{h_2(x', y_2)}. \tag{5.49}$$

From (5.48) and (5.49), it follows that

$$H_2(x, y) = \frac{f_2}{f(x, y)} h_2(x, y) \tag{5.50}$$

and therefore that

$$H_1(x, y) = \left(1 - \frac{f}{f_2}\right) H_2(x, y) \tag{5.51}$$

since $h_2 = H_2 - H_1$.

Equation (5.51) is the required second relation between $H_1$ and $H_2$ at $(x, y)$. This relation can also be obtained directly by evaluating the general functional dependencies (5.47) at the outcrop $y = y_2$, where $h_2 = H_2$. Because $B_2 = \gamma_2 H_2$,

$$\hat{Q}_2(\psi_2) = \tilde{Q}_2(B_2) = \frac{f_2}{h_2} = \frac{\gamma_2 f_2}{B_2} \quad \text{at } y = y_2, \tag{5.52}$$

and this implies that

$$Q_2 = \tilde{Q}_2(\gamma_2 H_2) = \frac{f_2}{H_2} \tag{5.53}$$

at any point $(x, y)$ connected to the outcrop by a streamline, from which (5.50) follows. The solution for $H_2$ can then be obtained from (5.45) and (5.51), which together yield

$$H_2(x, y) = \left[ \frac{h_E^2 + D_2^2(x, y)}{1 + \Gamma_1 \left(1 - \frac{f}{f_2}\right)^2} \right]^{1/2}. \tag{5.54}$$

With $H_2$ known, $H_1$ can in turn be obtained from (5.51).

This completes the solution in the region in which layer-2 trajectories can be traced back to the outcrop to obtain the relation (5.51). Because properties of the subsurface layer 2 in this region have been set through its upstream exposure to surface conditions in $y > y_2$, prior to its subduction beneath layer 1, this region may be considered a ventilated regime. The two-layer model provides the simplest example of this ventilated thermocline structure.

The solution (5.54) for $H_2$, the depth of the interface at the base of the moving fluid in the two-layer model, can be seen as a minor modification of the solution (5.36) for the depth $h_1$ of the interface at the base of the moving fluid in the one-layer model. The only difference is a factor that depends on the ratios of the reduced gravities and of the values of the Coriolis parameter at the local and outcrop latitudes. If $\gamma_2$ in (5.54) is set equal to $\gamma_1$ in (5.36), this factor is always greater than or equal to 1; thus the effect of additional upper-layer stratification is always to confine the Sverdrup transport closer to the surface. On the other hand, if the sum $\gamma_1 + \gamma_2$ in the two-layer model is set equal to $\gamma_1$ in (5.36), the depth of the moving fluid in the two-layer model will be greater than that in the one-layer model so that the effect of distributing the same total stratification over the two moving layers is a deepening of the wind-driven motion. In either case, the base of the moving layer will retain the qualitative half-bowl shape of the one-layer solution, and the curves of constant $H_2$, which determine the fluid trajectories in layer 2, will be roughly semicircular and concave toward the west or northwest.

The solution (5.51) for the upper-layer thickness $H_1$ in terms of $H_2$ shows that remarkably, the ratio of $H_1$ to the total depth $H_2$ of the moving fluid depends only on the local and outcrop latitudes. This ratio is independent of the reduced gravities $\gamma_1$ and $\gamma_2$ and of the structure of the forcing $W_E$. This is the result of the strong constraints placed on the subsurface flow by the conservation of potential vorticity and Bernoulli function along the layer-2 fluid trajectories.

The shape of these trajectories, the contours of constant $H_2$, is critical, because the solution (5.51) and (5.54) for $y < y_2$ has been derived under the assumption that the trajectory passing through the point $(x, y)$ also passes through the outcrop $y = y_2$ within the zonal extent $[x_W, x_E]$ of the basin. To determine the domain of validity of this solution, it is necessary to compute the trajectories that pass through the outcrop at the zonal extremes of the basin, $x = x_W$ and $x = x_E$.

### Shadow Zone

Consider first the trajectory leaving the point $(x_E, y_2)$, where the outcrop intersects the eastern boundary. Denote this trajectory by $(x, y) = [x_S(y), y]$. By (5.48), it follows that

$$H_2[x_S(y), y] = h_E \tag{5.55}$$

since $h_2(x_E, y_2) = h_E$. From the interior solution (5.54) for $H_2$, it follows that

$$\Gamma_1 \left(1 - \frac{f}{f_2}\right)^2 h_E^2 = D_2^2[x_S(y), y], \tag{5.56}$$

which is an implicit equation for $x_S(y)$. Suppose that $W_E$ is independent of $x$. Then (5.56) can be solved explicitly, yielding

$$x_S(y) = x_E + \frac{\beta \gamma_1 h_E^2}{2 f^2 W_E(y)} \left(1 - \frac{f}{f_2}\right)^2. \tag{5.57}$$

As $y \to y_2$, $f \to f_2$, and therefore $x_S(y) \to x_E$ as $y \to y_2$, by construction. Because $W_E$ is negative in the subtropical gyre, and all other quantities in the second term of (5.57) are positive, it follows that $x_S(y) < x_E$ for $y < y_2$, so the trajectory follows a course equatorward and westward from the eastern boundary into the subtropical gyre interior (Figure 5.6).

Between $x_S(y)$ and the eastern boundary, that is, for $(x, y)$ with $x_S(y) < x < x_E$ and $y < y_2$, layer-2 trajectories cannot be traced back to the outcrop latitude, and the solution (5.51) and (5.54) does not hold. Because $H_2 = h_E$ on both boundaries of this region, a consistent assumption is that $H_2 = h_E$ throughout the region. In this case, layer 2 is motionless throughout the region, which may be considered to be in the shadow of the eastern boundary; for this reason, it is called the *shadow zone*. All the Sverdrup transport in the shadow zone must be carried by layer 1, so the one-layer solution (5.36) applies, which yields

$$H_1(x, y) = \left[D_1^2(x, y)\right]^{1/2}, \quad x_S(y) < x < x_E \tag{5.58}$$

since layer 1 has vanishing depth at the eastern boundary. With (5.55) and (5.56), it is straightforward to verify that the solutions (5.51) and (5.58) for the upper-layer thickness are continuous at the shadow-zone boundary $x = x_S(y)$.

### Western Pool

Consider now the trajectory leaving the point $(x_W, y_2)$, where the outcrop intersects the western boundary. Denote this trajectory by $(x, y) = [x_P(y), y]$. By (5.48), it follows that

$$H_2[x_P(y), y] = H_{2W}, \tag{5.59}$$

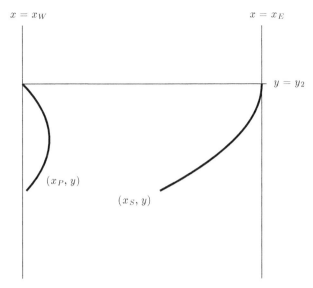

Figure 5.6. The shadow zone ($[x_S(y), y]$) and western pool ($[x_P(y), y]$) boundary trajectories in the subsurface layer 2. The origination point of $x_S(y)$ is located at $(x_E, y_2)$, where the outcrop latitude meets the eastern boundary, and $x_S(y)$ is tangent to the eastern boundary at that point. The origination point of $x_P(y)$ is located at $(x_W, y_2)$, where the outcrop latitude meets the western boundary; the figure shows the case in which the latitude of maximum downward Ekman pumping velocity is equatorward of $y = y_2$.

where $H_{2W} = H_2(x_W, y_2)$, and because $H_1(x, y_2) = 0$,

$$H_{2W} = \left[ h_E^2 + D_2^2(x_W, y_2) \right]^{1/2} .  \tag{5.60}$$

From (5.54), it then follows that

$$\left[ 1 + \Gamma_1 \left( 1 - \frac{f}{f_2} \right)^2 \right] H_{2W}^2 = h_E^2 + D_2^2[x_P(y), y],  \tag{5.61}$$

which is an implicit equation for $x_P(y)$. If, again, it is assumed that $W_E$ is independent of $x$, the solution for $x_P(y)$ may be obtained explicitly, yielding

$$x_P(y) = x_W - \frac{\beta y_2}{2 f^2 W_E(y)} \left\{ \Gamma_1 \left( 1 - \frac{f}{f_2} \right)^2 h_E^2 \right.$$

$$\left. + \left[ 1 + \Gamma_1 \left( 1 - \frac{f}{f_2} \right)^2 \right] \left[ D_2^2(x_W, y_2) - D_2^2(x_W, y) \right] \right\} .  \tag{5.62}$$

As $y \to y_2$, $f \to f_2$, and therefore $x_P(y) \to x_W$ as $y \to y_2$, by construction. The quantity $W_E$ is negative in the subtropical gyre, so if the term in the brackets in (5.62) is positive for $y < y_2$, the trajectory $x_P(y)$ follows a course equatorward and eastward from the western boundary into the subtropical gyre interior, whereas if it is negative, the trajectory $x_P(y)$ is directed westward and terminates immediately. At $y = y_2$, the term in brackets vanishes, and its derivative with respect to $y$ is equal to $-d D_2^2(x_W, y)/dy$, evaluated at $y = y_2$. Thus, if $d[f^2 W_E(y)]/dy > 0$ at $y = y_2$, $x_P(y)$ enters the interior, whereas if $d[f^2 W_E(y)]/dy < 0$ at $y = y_2$, $x_P(y)$ terminates immediately. These derivatives are typically dominated by $d W_E/dy$; thus the trajectory $x_P(y)$ will enter the interior if the outcrop latitude $y = y_2$ is sufficiently poleward of the latitude of maximum Ekman pumping $W_E$ and will not if the outcrop latitude is sufficiently equatorward of that latitude. In the former case, the trajectory $x = x_P(y)$ must have an approximately semicircular structure since, by (5.59), it is a contour of constant $H_2$, which, in turn, according to the preceding discussion, has the structure of a half-bowl deepening toward the west. In this case, the trajectory $x = x_P(y)$ separates the interior domain to its east in which the solution (5.51) and (5.54) is valid from an isolated western pool region in which layer-2 trajectories cannot be traced back to the layer-2 outcrop (Figure 5.6). As for the shadow zone adjacent to the eastern boundary, an additional assumption is required to obtain the solution in the western pool.

If it presumed that the subsurface layers of the western boundary current are adiabatic to first order so that they return fluid to the interior at the same density and temperature at which they receive it from the interior, then the western boundary current cannot provide a net source of fluid to subsurface layers in the western pool region. Because the bounding trajectory $x = x_P(y)$ is a streamline for layer-2 flow and there are no outcrops of layer 2 in the western pool, there is then no source at all for layer-2 fluid in the western pool. The natural conclusion is that there is then no layer-2 fluid in the western pool; instead, the pool must fill entirely with layer-1 fluid, which is continuously supplied by Ekman pumping at the surface. In this case, $H_2 = H_1$ in the western pool, and the Sverdrup balance (5.45) yields the solution

$$H_1(x, y) = H_2(x, y) = \left\{ \frac{\gamma_2}{\gamma_1 + \gamma_2} \left[ h_E^2 + D_2^2(x, y) \right] \right\}^{1/2} \tag{5.63}$$

in the western pool region $x < x_P(y)$. Note that $H_1$ and $H_2$ then have discontinuities at $x = x_P(y)$; some additional considerations, which will not be pursued here, are required to show that only small modifications to the trajectory $x = x_P(y)$ are sufficient to obtain a consistent theory that incorporates the dynamics at the pool boundary. The ventilated pool solution (5.63) suggests that the western pool region will be marked by weak stratification in the upper thermocline because the surface fluid fills the pool to the base of the wind-driven layer. In this way, it offers a simple theory for the existence of the weakly-stratified subtropical mode waters that are found in the poleward and western corners of all the ocean's subtropical gyres.

An alternative assumption is that the potential vorticity inside the pool region is homogenized and equal to its constant value on the bounding contour $x = x_P(y)$.

The theoretical motivation for this assumption will be deferred to Chapter 6; here only its consequences are considered. Along the bounding trajectory $x = x_P(y)$, $H_2 = H_{2W}$ and $Q_2 = f_2/H_{2W}$, and the assumption of potential vorticity homogenization therefore implies that

$$Q_2(x, y) = \frac{f_2}{H_{2W}}, \quad x_W < x < x_P(y). \tag{5.64}$$

Because, by definition,

$$Q_2 = f/h_2, \tag{5.65}$$

it follows from condition (5.64) that

$$H_2 - H_1 = \frac{f}{f_2} H_{2W}, \tag{5.66}$$

which, with the Sverdrup relation (5.45), yields a quadratic equation for $H_2$ inside the pool region:

$$(1 + \Gamma_1) H_2^2 - 2\Gamma_1 \frac{f}{f_2} H_{2W} H_2 = h_E^2 + D_2^2(x, y) - \Gamma_1 \frac{f^2}{f_2^2} H_{2W}^2. \tag{5.67}$$

The branch of the solution of the quadratic equation (5.67) should be chosen so that $H_2 = H_{2W}$ at $x = x_P(y)$. The layer-1 thickness may then be determined from (5.66). With homogenized potential vorticity in the western pool, the layer-2 thickness $h_2$ is a function of latitude only and decreases monotonically with latitude because $h_2 = (f/f_2)H_{2W}$ and $f_2$ and $H_{2W}$ are constants. The continuity of $Q_2$ at $x = x_P(y)$ ensures that both $h_2$ and $h_1$ are continuous across the edge of the pool. Thus, in contrast to the case of the ventilated pool, no special considerations are required at the pool boundary. The ventilated pool and potential vorticity homogenization theories for the structure of the western pool region are founded on contrasting assumptions: in the first, ventilation is presumed to dominate, expunging the subsurface fluid and filling the pool with ventilated fluid, whereas in the second, time-dependent mesoscale processes are presumed to sustain the recirculating subsurface layer and, in concert with the relatively rapid recirculation, to control the structure of the unventilated layer.

### *N-Layer Models*

The ventilated thermocline theory can be extended to an arbitrary number $N$ of moving layers by successively determining the functional relations (4.77) at the layer outcrops $\{y = y_j; \ y_j < y_{j+1}, \ j = 2, \dots, N\}$, as in (5.52)–(5.53) for $N = 2$. With sufficiently large $N$, essentially continuous representations of the meridional surface and vertical interior temperature and density gradients are possible in principle, replacing the idealized abrupt contrasts at the outcrop latitudes in models with small $N$. However, analytical solution of the problem for $N > 3$ is awkward, in part because each inter-section of the shadow zone or western pool trajectories with the overlying outcrops spawns new sets of bounding trajectories and corresponding sets of new subdomains, in which new relations between the potential vorticities and stream functions must be obtained.

The qualitative structure of the multilayer solutions remains similar to that of the two-layer solution. The ratio of the depths of the ventilated subsurface layers $j$ for any $j < N - 1$ to the total moving fluid depth $H_N$ for $N > 2$ remains dependent on the local and outcrop latitudes and independent of the imposed Ekman pumping $W_E$. In contrast to the result for $N = 2$, however, this ratio may depend on the reduced gravities $\gamma_j$. For example, with $N = 3$, the equation analogous to (5.51) for the thickness $h_2$ of layer $j = 2$ is

$$ h_2 = \frac{f}{f_2}\left(1 - \frac{f}{f_2}\right)\frac{1 + \Gamma_2\left(1 - \frac{f}{f_3}\right)}{1 + \Gamma_2\left(1 - \frac{f_2}{f_3}\right)}H_3. \tag{5.68} $$

In the ventilated pool, the multilayer solutions develop horizontal but not vertical gradients as the uniform pool of layer 1 fluid in the two-layer model is replaced by a nested set of annuli of successively outcropping layers, with layer 1 at the center and layer $N - 1$ at the outer annulus. The homogenized western pool regions of successively deeper layers are successively smaller and more tightly confined to the poleward, western corners of the subtropical gyre.

If the eastern boundary depth $h_E$ of the deepest moving layer is set to zero, so that there is no shadow zone, and the western boundary $x_W \to -\infty$, so that there is no western pool, the $N$-layer model may be solved for arbitrary $N$ by a recursive algorithm. Other important extensions beyond the theory presented here include representations of finite mixed-layer depths. In the preceding discussion, the upper-layer depth is assumed to decrease gradually to zero as the lower-layer outcrop is approached; with finite mixed-layer depths, a final portion of this decrease is allowed to occur abruptly at the outcrop latitude, introducing new effects into the dynamics of subduction of the density layers at the outcrops.

## 5.6 Internal Boundary Layer Theory

The adiabatic theory of the subtropical gyre thermocline, the one-layer and multilayer ventilated thermocline models, is motivated by the basic observation that the wind-driven Ekman vertical velocities near the surface are much larger than the vertical velocities induced by the turbulent vertical diffusion of heat. In all these models, however, the vertical velocity decreases monotonically with depth from the surface until it vanishes at the base of the wind-driven motion, the interface at the base of the deepest moving layer. The vanishing of the vertical velocity at this interface follows from the absence of any motion in the fluid below and from the continuity of the vertical velocity.

Near the base of the wind-driven motion, then, the magnitude of the small, diffusively driven vertical velocity will equal or exceed that of the wind-driven vertical velocity. In this region, the adiabatic theory will no longer be valid, and it is necessary to consider theories that incorporate a representation of the turbulent heat diffusion. The simplest such representation, in which the vertical turbulent flux of heat is modeled by down-gradient diffusion with constant thermal diffusivity $\kappa_v$, results in the modified thermodynamic equation (4.53).

Because it is the region at the base of the wind-driven motion that is of interest here, the representation of the upper thermocline is of secondary importance, and an adequate starting point is provided by the one-layer adiabatic model described in Section 5.3, which has the solution (5.36) for the thickness $h_1$ of the moving layer. The inclusion of the small vertical heat diffusion (4.53) can be anticipated to lead to the development of a diffusive internal boundary layer at the base of the moving layer, $z = -h_1(x, y)$, where the adiabatic theory represents the sharp vertical thermal gradient as a discontinuity across the layer interface (Figure 5.7).

Let the stretched internal boundary layer coordinate $\zeta$ be defined by

$$\zeta = [z + h_1(x, y)]/\delta_i, \tag{5.69}$$

where the internal boundary layer thickness scale $\delta_i$ will turn out to be equivalent to that in (5.31). The matching conditions on the temperature outside the internal boundary layer require that the boundary layer temperature approach the temperatures $T_1$ and $T_2$ of the adiabatic layers above and below the interface, respectively:

$$T(\zeta) \to T_1, \quad \zeta \to \infty \quad \text{and} \quad T(\zeta) \to T_2, \quad \zeta \to -\infty. \tag{5.70}$$

Similarly, the matching conditions on the vertical velocity require that the vertical velocity approach the wind-driven vertical velocity $w_1(z)$ in the moving layer above the interface; because the horizontal velocities are independent of depth in the moving layer, $w_1$ must be a linear function of $z$:

$$w(\zeta) \to w_1(z) = \left(\frac{z + h_1}{h_1}\right) w|_{z=0} = \delta_i \zeta \frac{W_E}{h_1}, \quad \zeta \to \infty. \tag{5.71}$$

In the layer beneath, the adiabatic motion vanishes, and the matching condition requires that the vertical velocity vanish to first order outside the boundary layer; a constant vertical velocity $w_\infty$ of order $\delta_i$, induced by the diffusion, is allowed:

$$w \to w_\infty = \delta_i w_0, \quad \zeta \to \infty, \tag{5.72}$$

where $w_0$ is a constant to be determined.

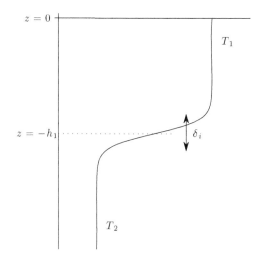

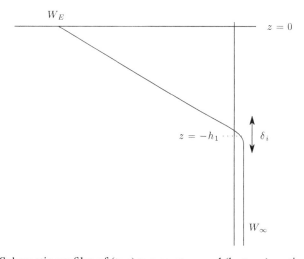

Figure 5.7. Schematic profiles of (top) temperature and (bottom) vertical velocity for the internal boundary layer thermocline model (5.80). The internal boundary layer has width $\delta_i$ and is located at $z = -h_1$. It is a transition layer between different, uniform temperatures $T_1$ above and $T_2$ below and between uniform Ekman-forced, vertical convergence $\partial w / \partial z < 0$ above and weak, diffusively driven upwelling below.

It is convenient to derive the equations for the internal boundary layer using the $M$ equation form (4.57) of the planetary geostrophic equations. A solution valid through the boundary layer may be sought in the form

$$M(x, y, z) = g\alpha_T \left\{ \frac{1}{2} z^2 T_2 + \delta_i^2 (T_1 - T_2) A[x, y, \zeta(x, y, z)] \right\}, \qquad (5.73)$$

where, for clarity, the dependence of $\zeta$ on $x$, $y$, and $z$, from (5.69), has been shown explicitly. The temperature $T$ is

$$T = \frac{1}{g\alpha_T}\frac{\partial^2 M}{\partial z^2} = T_2 + (T_1 - T_2)\frac{\partial^2 A}{\partial \zeta^2}, \qquad (5.74)$$

so in terms of $A$, the temperature matching conditions are

$$\frac{\partial^2 A}{\partial \zeta^2} \to 1, \quad \zeta \to \infty \quad \text{and} \quad \frac{\partial^2 A}{\partial \zeta^2} \to 0, \quad \zeta \to -\infty. \qquad (5.75)$$

The vertical velocity $w$ is

$$w = \frac{\beta}{f^2}\frac{\partial M}{\partial x} = \frac{\beta}{f^2}\delta_i g\alpha_T(T_1 - T_2)\left(\frac{\partial A}{\partial \zeta}\frac{\partial h_1}{\partial x} + \delta_i\frac{\partial A}{\partial x}\right), \qquad (5.76)$$

so the vertical velocity matching conditions are, to first order in $\delta_i$,

$$\frac{\partial A}{\partial \zeta} \to \zeta, \quad \zeta \to \infty \quad \text{and} \quad \frac{\partial A}{\partial \zeta} \to w_0, \quad \zeta \to -\infty. \qquad (5.77)$$

Here use has been made of the Sverdrup relation (5.35), with $\gamma_1 = g\alpha_T(T_1 - T_2)$.

The boundary layer equation for $A(x, y, \zeta)$ must be obtained by substituting (5.73) into the $M$ equation (4.57). The result is

$$f^{-1}\left[\frac{\partial h_1}{\partial x}\left(\frac{\partial^3 A}{\partial \zeta^2 \partial y}\frac{\partial^2 A}{\partial \zeta^2} - \frac{\partial^3 A}{\partial \zeta^3}\frac{\partial^2 A}{\partial \zeta \partial y}\right) - \frac{\partial h_1}{\partial y}\left(\frac{\partial^3 A}{\partial \zeta^2 \partial x}\frac{\partial^2 A}{\partial \zeta^2} - \frac{\partial^3 A}{\partial \zeta^3}\frac{\partial^2 A}{\partial \zeta \partial x}\right)\right]$$

$$+ \beta f^{-2}\frac{\partial h_1}{\partial x}\frac{\partial A}{\partial \zeta}\frac{\partial^3 A}{\partial \zeta^3} = \delta_i^{-2}\gamma_1^{-1}\kappa_v\frac{\partial^4 A}{\partial \zeta^4} + \mathcal{O}(\delta_i). \qquad (5.78)$$

So that the diffusive term enters the balance at first order, it is necessary to choose

$$\delta_i \propto \kappa_v^{1/2}. \qquad (5.79)$$

Consequently, $\delta_i$ may be identified with the internal boundary layer scale (5.31), and the vertical velocity $\delta_i w_0$ induced beneath the interface by the diffusion will also be of order $\kappa_v^{1/2}$.

Because the matching conditions (5.75) and (5.77) are independent of $x$ and $y$, it might have been hoped that the boundary layer equation (5.78) would reduce to a one-dimensional equation, with derivatives only in the boundary layer coordinate $\zeta$. Unfortunately, this is not the case: because the vertical velocity decreases toward zero in the boundary layer, the small horizontal advective terms remain of the same order as the vertical advective term, and the thermodynamic balance does not automatically reduce to the vertical advective-diffusive balance $w\,\partial T/\partial z = \kappa_v \partial^2 T/\partial z^2$ as $\kappa_v \to 0$. Consistent with the matching conditions, solutions can nonetheless be sought that depend only on $\zeta$ by making the substitution $A(x, y, \zeta) = G(\zeta)$ in (5.78). The result is a nonlinear ordinary differential equation:

$$\frac{\delta_i^2}{\kappa_v} \frac{W_E}{h_1} \frac{dG}{d\zeta} \frac{d^3G}{d\zeta^3} = \frac{d^4G}{d\zeta^4}, \tag{5.80}$$

where, again, (5.35) has been used to simplify the coefficient term. The equation (5.80) and the assumed form of the solution are consistent with (5.78) only if the ratio $W_E/h_1$ is independent of $x$ and $y$, and from (5.36), it can be seen that this will not be true in general. Special forms of the forcing can be chosen to satisfy this condition to obtain a consistent one-dimensional equation that can be solved as an illustrative example. For example, let $h_E = 0$ and $W_E = C_w^2 f^2 (x - x_E)/(\beta \gamma_1)$, where $C_w$ is a constant, so that the downward Ekman pumping increases linearly westward from the eastern boundary and quadratically with latitude. Then, from (5.36),

$$\frac{W_E}{h_1} = C_w, \tag{5.81}$$

so with the choice

$$\delta_i = \left( \frac{\kappa_v h_1}{W_E} \right)^{1/2} = \left( \frac{\kappa_v}{C_w} \right)^{1/2} \tag{5.82}$$

and the substitution

$$F(\zeta) = \frac{dG}{d\zeta}, \tag{5.83}$$

Equation (5.80) reduces to

$$F \frac{d^2F}{d\zeta^2} = \frac{d^3F}{d\zeta^3}. \tag{5.84}$$

where prime denotes differentiation with respect to the argument. The corresponding matching conditions are

$$F(\zeta) \to \zeta, \quad \zeta \to \infty \quad \text{and} \quad \frac{dF}{d\zeta}(\zeta) \to 0, \quad \zeta \to -\infty, \tag{5.85}$$

while the induced vertical velocity is determined by the constant limit of $F(\zeta)$ as $\zeta \to -\infty$. This scaled problem depends on no parameters, and numerical solution shows that

$$F(\zeta) \to 0.876, \quad \zeta \to -\infty, \tag{5.86}$$

so that the induced vertical velocity $w_\infty$ beneath the diffusive internal boundary layer is

$$w_\infty = w(\zeta \to -\infty) = 0.876 \, \delta_i \frac{W_E}{h_1} = 0.876 \, \delta_i C_w = 0.876 \, (\kappa_v C_w)^{1/2}. \tag{5.87}$$

Thus the diffusively driven upwelling is proportional to the product of the thickness of the internal boundary layer and the vertical convergence of the adiabatic flow in the upper layer. Because $h_1$ scales with the advective depth $D_a$, this result for the upwelling velocity is consistent with the scaling estimate (5.31) for both the square root dependence on the vertical diffusivity and the weaker, one-fourth-power dependence on the wind forcing $W_E$.

## 5.7 Diffusively Driven Circulation

The internal boundary layer theory provides an explicit example of the generation of large-scale vertical motion by small-scale turbulent diffusion in the presence of a mean vertical temperature gradient. Such a driving of large-scale motion by vertical turbulent diffusive fluxes was anticipated from the energy equation (5.12). In the case of the rectangular, single-hemisphere basin, the general structure of the resulting large-scale, diffusively driven deep circulation flow can be inferred from basic dynamical considerations.

Suppose that the deep motionless layer $j = N + 1$ beneath the wind-driven fluid in an adiabatic model of the subtropical gyre extends to the seafloor in an ocean basin of uniform depth $H_0$. On the basis of the scalings (5.26) or (5.31), it can be anticipated that in the presence of weak vertical turbulent heat diffusion, $\kappa_v > 0$ in (4.53), a vertical velocity $w_\infty$ proportional to $\kappa_v^{1/3}$ or $\kappa_v^{1/2}$ will be induced at the top of this otherwise motionless layer. In the case of the internal boundary layer equation (5.84), the associated vertical velocity is given explicitly by (5.87).

By the Sverdrup transport balance, it follows that the diffusively driven vertical velocity will in turn induce a geostrophic meridional flow in the deep layer:

$$\beta V_{N+1} = f w_\infty, \tag{5.88}$$

where $V_{N+1} = h_{N+1} v_{N+1} = (H_0 - H_N) v_{N+1}$ is the vertically integrated geostrophic meridional transport in the deep layer. Because $w_\infty > 0$, this diffusively driven meridional transport will be directed poleward. Thus in the presence of weak diffusion, there will be a broad, weak, poleward geostrophic flow beneath the main thermocline in this model. The magnitude of the corresponding zonally integrated deep poleward transport may be estimated as $V_{N+1} L \approx (\kappa_v D_a / W_E)^{1/2} L^2 \approx 5 \times 10^6$ m$^3$ s$^{-1}$ = 5 Sv, or roughly one-fifth the transport of the wind-driven subtropical upper ocean gyre.

The deep, cold fluid that upwells with velocity $w_\infty$ into the base of the main thermocline gains heat through the vertical divergence of the vertical turbulent diffusive heat flux. A source of deep, cold fluid is required to balance this divergence. This cold fluid must form at high latitudes, where surface cooling by heat exchange with the overlying atmosphere can occur. Thus the source of deep fluid that balances the upwelling must lie at high latitudes. However, the Sverdrup interior flow (5.88) is itself directed toward high latitudes. The constraints of the large-scale interior vorticity balance imply that the diffusively driven upwelling divergence and this interior flow must be supported by a deep western boundary current that can transport the deep, cold fluid from high latitudes to midlatitudes, where, in turn, it may leave the boundary and flow zonally into the interior.

## 5.8 Meridional Overturning Cells

With no-normal-flow conditions at the eastern and western boundaries, the zonal integral of the incompressibility condition (2.104) yields a two-dimensional incompressibility condition for the zonally integrated meridional and vertical flow:

$$
0 = \int_{x_W}^{x_E} \nabla \cdot \mathbf{u} \, dx
$$
$$
= \frac{\partial}{\partial y} \int_{x_W}^{x_E} v \, dx + \frac{\partial}{\partial z} \int_{x_W}^{x_E} w \, dx. \tag{5.89}
$$

It follows that a stream function $\Psi_m(y, z)$ may be introduced for the zonally integrated flow so that

$$
\int_{x_W}^{x_E} v \, dx = -\frac{\partial \Psi_m}{\partial z}, \qquad \int_{x_W}^{x_E} w \, dx = \frac{\partial \Psi_m}{\partial y}. \tag{5.90}
$$

The circulations described by sets of closed contours around extrema of this meridional overturning stream function $\Psi_m$ are the meridional overturning cells. With the sign convention in (5.90), cells around $\Psi$ maxima have northward flow above the maximum point and southward flow beneath, whereas cells around $\Psi$ minima have southward flow above and northward flow beneath.

In the single-basin circulation patterns analyzed in the preceding sections, there are three main meridional overturning cells: a pair of shallow, counterrotating, wind-driven cells above the main thermocline and a third, abyssal cell beneath the main thermocline. The wind-driven cells describe the northward and southward surface Ekman transports toward the central latitudes of the subtropical gyre, with subsurface return flows in the ventilated thermocline and the western boundary current. The abyssal cell describes the diffusion-driven interior warming and upwelling of cold water and the return flow supported by surface cooling, convective adjustment, and sinking at subpolar latitudes, followed by southward transport in the deep western boundary current.

## 5.9  Summary

Scaling estimates for the depth and thickness of the wind-driven and diffusively driven subtropical gyre thermocline, and the associated vertical velocities, show that the dominant thermodynamic balance in the upper midlatitude thermocline is advective and adiabatic. The consequences of this advective balance can be conveniently explored with single-layer and multilayer adiabatic models. At the base of the wind-driven motion, the associated vertical velocity becomes small, and diffusive effects must be considered. This diffusion supports an internal boundary layer at the base of the wind-driven motion and induces a weak upwelling of the deep, underlying fluid.

From these results, it can be inferred that the temperature contrasts across the two components of the thermocline, the adiabatic, ventilated thermocline above and the internal boundary layer at its base, are controlled by the thermal contrasts across the subtropical and subpolar gyres, respectively (Figure 5.8). The coldest fluid in the ventilated thermocline is that injected into the interior at the subtropical-subpolar gyre boundary, which forms the deepest ventilated layers, whereas the warmest fluid is that injected at the farthest southern latitudes of the gyre. The abyss, on the other hand, is filled with the coldest available fluid, the temperature of which is set by the coldest surface temperatures in the subpolar gyre; the internal boundary layer forms between this cold fluid and the deepest ventilated layer, which is just that injected from the surface at the subtropical-subpolar gyre boundary.

The circulation described by these analytical models of the subtropical gyre thermocline must be closed by boundary currents, in which the dominant motions and transports are in the western boundary layer, and by an assumed subpolar gyre circulation that includes Ekman upwelling in the interior, surface cooling and convection, and sinking and southward export of cooled fluid. The wind-driven cyclonic circulation of the depth-integrated motion of the subpolar gyre can be easily obtained from the Sverdrup transport balance (3.30), but the importance of diabatic processes and subpolar boundary currents and the control of incoming fluid properties by the western boundary current make quantitative analytical modeling of the stratified subtropical gyre circulation difficult.

## 5.10  Notes

The advective and advective-diffusive boundary-layer scalings for the thermocline were derived by Welander (1971). The one-layer reduced-gravity model of the subtropical gyre was developed independently by Parsons (1969) and Veronis (1973). The multilayer, ventilated thermocline theory of the subtropical upper thermocline is from Luyten et al. (1983). A continuously stratified ventilated thermocline theory has been worked out by Huang (1988). The diffusive main thermocline and the associated diffusively driven, abyssal meridional overturning cell were studied by Stommel and Arons (1960) and Stommel and Webster (1962);

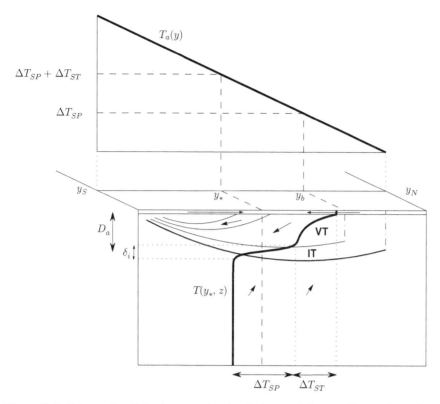

Figure 5.8. Schematic of the large-scale circulation and thermocline regimes in a simply connected, single-hemisphere ocean basin driven by surface wind stress, weak turbulent diffusion, and a meridional surface thermal gradient. A representative ocean vertical temperature profile $T(y_*, z)$ at a central subtropical latitude $y = y_*$ is superimposed on a midbasin meridional section of ocean thermocline isotherms. The surface or atmospheric temperature $T_a(y)$ above decreases monotonically from its warmest value at the extreme tropical latitude $y_S$ to its coldest value at the extreme subpolar latitude $y_N$. Circulation in the meridional plane is indicated by arrows on the section. The convergent lateral surface Ekman transport resulting from tropical easterly and midlatitude westerly wind forcing is indicated by opposing horizontal arrows in the ocean-surface layer. The ventilated thermocline (VT) is an advective regime with characteristic thickness $D_a$ and equatorward Sverdrup flow. The characteristic vertical thermal contrast across the VT regime is the difference $\Delta T_{ST} = T_a(y_*) - T_a(y_b)$ between the local surface temperature and the surface temperature at the subtropical-subpolar gyre boundary $y = y_b$. The internal thermocline (IT) at the base of the VT is a diffusive internal boundary layer regime, with characteristic thickness $\delta_i$. The characteristic vertical thermal contrast across the IT regime is the surface difference $\Delta T_{SP} = T_a(y_b) - T_a(y_N)$ between the surface temperature at the subtropical-subpolar gyre boundary $y = y_b$ and the coldest surface temperatures at the extreme subpolar latitude $y_N$. The induced deep upwelling into the IT drives poleward abyssal Sverdrup flow. (Redrawn from Samelson and Vallis [1997]. Used with permission.)

the latter studied a similarity solution related to the discussion in Section 5.6, for which the reduced equation (5.84) also applies and was obtained and solved by Young and Ierley (1986). Salmon (1990) introduced the concept of the internal boundary layer as a conceptual model for the subtropical main thermocline. The ventilated-pool theory for subtropical mode water was developed by Dewar et al. (2005), who include a more detailed treatment of conditions at the boundary of the pool.

# 6

# Eddy-Driven Subsurface Motion

## 6.1 Homogenization of Scalars in Recirculating Flow

In the preceding chapter, some theoretical results for the subtropical gyre circulation were obtained that relied on the assumption of homogenized potential vorticity within closed recirculation contours. The motivation for this assumption is explored in this chapter. The basic mechanism at work can be most easily understood in the simpler case of homogenization of a passive scalar in recirculating two-dimensional flow.

Consider the advection-diffusion equation for a passive scalar $C$ in a steady two-dimensional flow given by a stream function $\psi(x, y)$,

$$\mathbf{u}_h \cdot \nabla_h C = K_C \nabla_h^2 C, \tag{6.1}$$

where $\mathbf{u}_h = (-\partial\psi/\partial y, \partial\psi/\partial x)$ and $K_C$ is the diffusivity of the scalar. Suppose that $K_C$ is positive but small, that is,

$$0 < \frac{K_C}{UL} = \frac{1}{\mathrm{Pe}_C} \ll 1, \tag{6.2}$$

where $U$ and $L$ are representative velocity and length scales for the flow and $\mathrm{Pe}_C$ is a Peclét number for the passive scalar $C$. Then, to first order in $K_C$ (or, more formally, in $\mathrm{Pe}_C^{-1}$), the scalar is conserved along streamlines so that

$$C = \hat{C}(\psi) + \mathcal{O}(K_C). \tag{6.3}$$

Suppose further that there is a region of flow enclosed by a closed streamline $\Gamma$ of $\psi$ (Figure 6.1), and integrate the equation (6.1) over the area $A_\Gamma$ enclosed by $\Gamma$:

$$\int_{A_\Gamma} \mathbf{u}_h \cdot \nabla_h C \, dx \, dy = K_C \int_{A_\Gamma} \nabla_h^2 C \, dx \, dy. \tag{6.4}$$

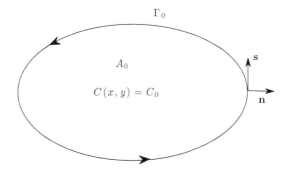

Figure 6.1. Streamline geometry for passive scalar homogenization in steady two-dimensional flow. The region $A_0$ consists of closed streamlines, with outermost closed streamline $\Gamma_0$. For sufficiently large Peclét number, the scalar concentration $C(x, y)$ will be homogenized throughout the recirculation region $A_0$, and equal to its value $C_0$ on $\Gamma_0$. The vectors **n** and **s** are the unit normal and tangent, respectively, to $\Gamma_0$.

Then

$$
\int_{A_\Gamma} \mathbf{u}_h \cdot \nabla_h C \, dx \, dy = \int_{A_\Gamma} \nabla_h \cdot (\mathbf{u}_h C) \, dx \, dy
$$

$$
= \oint_\Gamma \mathbf{u}_h C \cdot \mathbf{n} \, ds
$$

$$
= \pm \oint_\Gamma \mathbf{k} \times \nabla_h \psi \cdot \frac{\nabla_h \psi}{|\nabla_h \psi|} C \, ds = 0, \qquad (6.5)
$$

where the last equality follows because the velocity is tangent to the streamline and therefore normal to the outward unit normal **n** to the contour $\Gamma$. Thus there is no advection of the passive scalar into or out of the region enclosed by the streamline $\Gamma$. With (6.3), it follows also that

$$
0 = K_C \int_{A_\Gamma} \nabla_h^2 C \, dx \, dy,
$$

$$
= K_C \oint_\Gamma \nabla_h C \cdot \mathbf{n} \, ds,
$$

$$
= \pm K_C \frac{d\hat{C}}{d\psi} \oint_\Gamma \nabla_h \psi \cdot \frac{\nabla_h \psi}{|\nabla_h \psi|} \, ds,
$$

$$
= \pm K_C \frac{d\hat{C}}{d\psi} \oint_\Gamma |\nabla_h \psi| \, ds \qquad (6.6)
$$

to first order in $K_C$. The latter integral is positive definite because the velocity must be nonzero on the streamline $\Gamma$, and consequently,

$$
\frac{d\hat{C}}{d\psi} = 0 \quad \text{on} \quad \Gamma. \qquad (6.7)
$$

Because Equation (6.7) is valid to first order in $K_C$ for any closed streamline $\Gamma$, it follows that to first order in $K_C$, the scalar $C$ must be constant throughout the full region $A_0$ of recirculating flow and equal to its value $C_0$ on the outermost closed streamline $\Gamma_0$ (Figure 6.1):

$$C(x, y) = C_0 \quad \text{for all} \quad (x, y) \in A_0. \tag{6.8}$$

The weak-diffusion condition (6.2) requires that fluid parcels make many circuits around a closed streamline $\Gamma$ before diffusion has a substantial effect. Under this condition (6.2), a passive scalar in steady two-dimensional flow will be homogenized within any region of closed streamlines.

## 6.2 Potential Vorticity Homogenization: Quasi-Geostrophic Theory

### *A Quasi-Geostrophic Approximation*

The general argument of the preceding section applies to a passive scalar, the distribution of which has, by definition, no effect on the imposed flow field given by the stream function $\psi$. If the advected scalar is the potential vorticity, however, the flow field and the distribution of the scalar field will instead be dynamically coupled. To determine whether a similar homogenization of the dynamical scalar potential vorticity can occur under qualitatively similar conditions of recirculation and weak scalar diffusion, it is necessary to solve the corresponding set of dynamical equations, with the passive scalar homogenization result as motivation and conceptual guidance.

Consider first a simplification of the two-layer planetary geostrophic equations for steady flow, in which the departures of layer thickness from constant values are small,

$$H_j = H_{j0} + \eta_j(x, y), \quad |\eta_j| \ll H_{j0} = \text{constant}, \quad j = \{1, 2\}, \tag{6.9}$$

and the meridional extent $\Delta y$ of the flow is restricted so that

$$\Delta y / L = \frac{\beta \Delta y}{f_0} \ll 1. \tag{6.10}$$

The corresponding linearized layer-$j$ potential vorticities are then $Q_j \approx f_0 / h_{j0} + q_j$, where

$$q_1 = \beta_1 y - \frac{f_0}{h_{10}^2} \eta_1 \tag{6.11}$$

$$q_2 = \beta_2 y - \frac{f_0}{h_{20}^2} (\eta_2 - \eta_1), \tag{6.12}$$

with $h_{10} = H_{10}$, $h_{20} = H_{20} - H_{10}$, and $\beta_j = \beta / h_{j0}$ for $j = \{1, 2\}$. In addition, the horizontal velocity in each layer can then be obtained approximately from a stream function,

$$(u_j, v_j) = \left( -\frac{\partial \phi_j}{\partial y}, \frac{\partial \phi_j}{\partial x} \right), \quad j = \{1, 2\}, \tag{6.13}$$

where

$$\phi_1 = \frac{\gamma_2}{f_0} \eta_2 + \frac{\gamma_1}{f_0} \eta_1, \quad \phi_2 = \frac{\gamma_2}{f_0} \eta_2, \tag{6.14}$$

and thus, to the same order of approximation,

$$q_1 = \beta_1 y - \frac{f_0^2}{\gamma_1 h_{10}^2}(\phi_1 - \phi_2) \tag{6.15a}$$

$$q_2 = \beta_2 y - \frac{f_0^2}{\gamma_2 h_{20}^2}\phi_2 - \frac{f_0^2}{\gamma_1 h_{20}^2}(\phi_2 - \phi_1). \tag{6.15b}$$

With these restrictions on the amplitude of thickness devations and the meridional extent of the flow, the resulting simplified planetary geostrophic equations are equivalent to the long-wave limit of the quasi-geostrophic equations.

By analogy with (6.1), suppose that a small dissipative term $\mathcal{D}_2$ is added to the equation for the rate of change of layer 2 potential vorticity in (5.46), which, with (6.12) and (6.13), can be written as

$$\mathbf{u}_2 \cdot \nabla_h q_2 = \beta_2 \frac{\partial \phi_2}{\partial x} + \frac{f_0^2}{\gamma_1 h_{20}^2} J(\phi_2, \phi_1) = \mathcal{D}_2 \tag{6.16}$$

because $J(\phi_2, \phi_2) \equiv 0$. In view of the smallness of the thickness disturbances (6.9), the Sverdrup transport balance (5.45) takes the linearized form

$$h_{10}\phi_1 + h_{20}\phi_2 = H_{20}\phi_B = \frac{f_0}{\beta} \int_{x_E}^{x} W_E(x', y) \, dx', \tag{6.17}$$

where $\phi_B$ is the barotropic stream function. This may be solved for $\phi_1$,

$$\phi_1 = \frac{H_{20}}{h_{10}}\phi_B - \frac{h_{20}}{h_{10}}\phi_2, \tag{6.18}$$

so that (6.16) may be written as

$$J(\phi_2, \hat{q}_2) = \mathcal{D}_2, \tag{6.19}$$

where

$$\hat{q}_2 = \beta_2 y + \frac{\hat{F}}{h_{20}}\phi_B, \quad \hat{F} = \frac{f_0^2 H_{20}}{\gamma_1 h_{10} h_{20}}. \tag{6.20}$$

This reformulation of (6.16) has achieved a fundamental simplication: because $\phi_B$ can be computed from the imposed forcing $W_E(x, y)$, (6.19) is a linear equation for $\phi_2$. In addition, because $\mathcal{D}_2$ is small by assumption, it follows that

$$\phi_2 = \hat{\Phi}_2(\hat{q}_2) + \mathcal{O}(\mathcal{D}_2). \tag{6.21}$$

Because, then, by (6.12) and (6.18)–(6.21),

$$q_2 = \hat{q}_2 - \frac{\hat{F} + F_2}{h_{20}} \hat{\Phi}_2(\hat{q}_2) + \mathcal{O}(\mathcal{D}_2), \tag{6.22}$$

where $F_2 = f_0^2/(\gamma_2 h_{20})$, it follows also that

$$\phi_2 = \Phi_2(q_2) + \mathcal{O}(\mathcal{D}_2). \tag{6.23}$$

Thus the stream function $\phi_2$ and potential vorticity $q_2$ are both constant on the geostrophic contours, the contours of constant $\hat{q}_2$. An analogy with the passive-scalar homogenization theory can then be anticipated if, in a given flow, there are geostrophic contours that are closed.

### A Mid-Ocean Gyre Example

It is useful here to consider a specific, though artificial, example of imposed Ekman pumping for which explicit solutions of these equations may be obtained. Let

$$W_E = \begin{cases} -\alpha x, & r \leq r_1, \\ 0, & r > r_1, \end{cases} \tag{6.24}$$

where $r = (x^2 + y^2)^{1/2}$, $\alpha = W_0/r_1$, $W_0$ and $r_1$ are given constants, and, to simplify notation for this example, the origin of the $y$ coordinate has been shifted to the central latitude $y_0$, where $f = f_0$. Then the barotropic stream function $\phi_B$ is given by

$$\phi_B(x, y) = \frac{\alpha f_0}{2\beta H_{20}} \left(r_1^2 - x^2 - y^2\right), \quad r \leq r_1, \tag{6.25}$$

and $\phi_B = 0$ for $r > r_1$. From (6.20),

$$\hat{q}_2(x, y) = \frac{\beta_2}{2y_1} \left[r_1^2 + y_1^2 - x^2 - (y - y_1)^2\right], \quad r \leq r_1, \tag{6.26}$$

and

$$\hat{q}_2(x, y) = \beta_2 y, \quad r > r_1, \tag{6.27}$$

where

$$y_1 = \frac{\beta^2 H_{20}}{\alpha f_0 \hat{F}}. \tag{6.28}$$

The geostrophic contours are thus arcs of circles for $r \leq r_1$, centered at $(0, y_1)$, and zonal lines $y = $ constant for $r > r_1$. Closed geostrophic contours will exist if $y_1 < r_1$, that is, if

$$W_0 > W_{0\mathrm{crit}} = \frac{\beta^2 H_{20}}{f_0 \hat{F}}. \tag{6.29}$$

The condition $y_1 = r_1$, or $W_0 = W_{0\mathrm{crit}}$ in (6.29), has an appealing physical interpretation: it is the forcing amplitude for which the zonal barotropic Sverdrup velocity $u_B = -\partial\phi_B/\partial y$ at $y = r_1$ is equal and opposite to the long baroclinic two-layer Rossby wave speed $c_R = -\beta/\hat{F}$:

$$u_B(0, r_1) = -\frac{\partial\phi_B}{\partial y}(0, r_1) = \frac{\alpha f_0 r_1}{\beta H_{20}} = \frac{\beta}{\hat{F}} = -c_R. \tag{6.30}$$

The latter wave speed can be derived from the linearized, time-dependent form of (4.72) with $W_E = 0$,

$$\frac{\partial q_j}{\partial t} + \beta_j \frac{\partial \phi_j}{\partial x} = 0, \quad j = \{1, 2\}, \tag{6.31}$$

which, with (6.11)–(6.14), yields

$$\frac{\partial}{\partial t}(\phi_1 - \phi_2) - \frac{\beta}{\hat{F}}\frac{\partial}{\partial x}(\phi_1 - \phi_2) = 0, \tag{6.32}$$

provided that the base of layer 2 is treated as a rigid boundary so that the second term in (6.12), which arises from deformations of the lower interface, is neglected, while the terms in (6.14) proportional to $\eta_2$ are retained as bottom pressure perturbations. Thus it is when the zonal Sverdrup flow becomes sufficiently strong to halt the westward propagation of long baroclinic Rossby waves that closed geostrophic contours appear in layer 2. The region of closed contours will be circular, centered at $(0, y_1)$, with radius $r_1 - y_1$. The radius of the region will always be less than or equal to $r_1$ because $y_1 \geq 0$, and the outermost contour of the region will always be tangent to the circle $r = r_1$ at the point $(0, r_1)$.

The potential vorticity equation (6.16) in layer 2 may now be integrated over the area $A_\Gamma$ enclosed by any closed geostrophic contour $\Gamma$, with the result that

$$\int_{A_\Gamma} \mathcal{D}_2 \, dx \, dy = \int_{A_\Gamma} \mathbf{u}_2 \cdot \nabla_h q_2 \, dx \, dy$$

$$= \int_{A_\Gamma} \nabla_h \cdot (\mathbf{u}_2 q_2) \, dx \, dy$$

$$= \oint_\Gamma \mathbf{u}_2 q_2 \cdot \mathbf{n} \, ds$$

$$= \pm Q_2(\Gamma) \oint_\Gamma \mathbf{k} \times \nabla_h \phi_2 \cdot \frac{\nabla_h \phi_2}{|\nabla_h \phi_2|} \, ds = 0, \tag{6.33}$$

where $Q_2(\Gamma)$ is the value of $q_2$ on the geostrophic contour $\Gamma$ and where the last two equalities follow because both $q_2$ and $\phi_2$ are constant on $\Gamma$.

### *Potential Vorticity Diffusion*

The solution for $q_2$ inside the region of closed geostrophic contours depends on the form of the dissipation $\mathcal{D}_2$. Suppose first that the dissipation results from the divergence of a down-gradient flux of potential vorticity,

$$\mathcal{D}_2 = K_q \nabla_h^2 q_2, \tag{6.34}$$

where $K_q$ is a constant that may be interpreted as an eddy diffusion coefficient. Then, exactly as for the passive scalar, it follows that

$$\begin{aligned}
0 &= K_q \int_{A_\Gamma} \nabla_h^2 q_2 \, dx \, dy \\
&= K_q \oint_\Gamma \nabla_h q_2 \cdot \mathbf{n} \, ds \\
&= \pm K_q \frac{d\hat{Q}_2}{d\phi_2} \oint_\Gamma |\nabla_h \phi_2| \, ds
\end{aligned} \tag{6.35}$$

to first order in $K_q$, where $Q_2(\phi_2)$ is the inverse of $\Phi_2(q_2)$. The latter integral is positive definite because the velocity must be nonzero on the streamline $\Gamma$, and consequently,

$$\frac{dQ_2}{d\phi_2} = 0 \quad \text{on} \quad \Gamma. \tag{6.36}$$

Because Equation (6.36) is valid to first order in $K_q$ for any closed streamline $\Gamma$, it follows that to first order in $K_q$, the potential vorticity $q_2$ must be constant and equal to its value $Q_2(\Gamma_0)$ on the outermost closed streamline $\Gamma_0$ of the region $A_0$ of recirculating flow:

$$q_2(x, y) = Q_2(\Gamma_0) \quad \text{for all} \quad (x, y) \in A_0. \tag{6.37}$$

Thus, when the dissipation $\mathcal{D}_2$ has the diffusive form (6.34), potential vorticity in layer 2 is homogenized throughout the circular region of closed streamlines centered at $(0, y_1)$ with radius $r_1 - y_1$.

Because it follows from (6.22) and (6.26) that $Q_2(\Gamma_0) = \beta_2 r_1$, the complete solution in layer 2 is

$$q_2(x, y) = \beta_2 r_1 \tag{6.38}$$

$$\phi_2(x, y) = \frac{\beta(y - r_1) + \hat{F}\phi_B(x, y)}{\hat{F} + F_2} \tag{6.39}$$

in the region $A_0$ of closed geostrophic contours, and

$$q_2(x, y) = \hat{q}_2(x, y) = \beta_2 y + \frac{\hat{F}}{h_{20}} \phi_B(x, y) \tag{6.40}$$

$$\phi_2(x, y) = 0 \tag{6.41}$$

outside $A_0$. Note that $\Phi_2 \to 0$ as $y \to 0$ for $x = 0$, so $\phi_2$ is continuous at $\Gamma_0$, and that (6.40) and (6.41) hold both in the part of the region $r < r_1$ that is outside $A_0$, where $\phi_B \neq 0$, and in $r > r_1$, where $\phi_B \equiv 0$ and $q_2 = \beta_2 y$. Thus the potential vorticity contours in layer 2 are distorted away from lines of constant latitude in the region of forcing, even if there is no motion in layer 2, owing to the deformation of the interface at the base of layer 1. It is when this interface become sufficiently deformed by the forcing, according to the criterion (6.29), that closed geostrophic contours form and motion is induced in the subsurface layer 2.

From the form of the analytical solution (6.38)–(6.41), as well as the general underlying requirement that closed geostrophic contours exist, it can be inferred that the scale for layer-thickness variations $\Delta \eta$ associated with potential vorticity homogenization must be that which gives potential vorticity variations that are comparable to the variations $\beta_j \Delta y$ in layer-$j$ background potential vorticity over the meridional scale $\Delta y$. For $\Delta y = 500$ km and $h_{j0} = 500$ m, this estimate gives $\Delta \eta \approx 50$ m. This is consistent with the linearization (6.9) and means also that $\Delta \eta \ll D_a$, where $D_a$ is the advective scale (5.20). If larger horizontal scales are considered, the predicted thickness variations will be proportionally larger, but the restriction (6.10) may be violated.

### *Vorticity Drag*

Suppose now that instead of the diffusive form (6.34), the dissipation takes the form of vorticity drag between layers 1 and 2:

$$\mathcal{D}_2 = -R_2(\zeta_2 - \zeta_1), \quad \zeta_j = \frac{\partial v_j}{\partial x} - \frac{\partial u_j}{\partial y}, \quad j = \{1, 2\}, \tag{6.42}$$

where $R_2$ is a constant drag coefficient. Then the integral (6.33) yields instead

$$0 = -R_2 \int_{A_\Gamma} (\zeta_2 - \zeta_1) \, dx \, dy$$

$$= -R_2 \oint_\Gamma (\mathbf{u}_2 - \mathbf{u}_1) \cdot \mathbf{s} \, ds, \tag{6.43}$$

where $\mathbf{s}$ is the unit tangent to the contour $\Gamma$. From (6.17), it then follows that

$$\oint_\Gamma \mathbf{u}_2 \cdot \mathbf{s} \, ds = \oint_\Gamma \mathbf{u}_B \cdot \mathbf{s} \, ds, \tag{6.44}$$

where $\mathbf{u}_B = (-\partial \phi_B/\partial y, \partial \phi_B/\partial x)$ is the barotropic velocity. Because, also,

$$\mathbf{u}_2 = \mathbf{k} \times \nabla_h \phi_2 = \frac{d\Phi_2}{d\hat{q}_2} \mathbf{k} \times \nabla_h \hat{q}_2, \qquad (6.45)$$

it follows that

$$\oint_\Gamma \mathbf{u}_2 \cdot \mathbf{s}\,ds = \frac{d\Phi_2}{d\hat{q}_2} \oint_\Gamma \left( \frac{\hat{F}}{h_{20}} \mathbf{u}_B + \beta_2 \mathbf{j} \right) \cdot \mathbf{s}\,ds = \frac{\hat{F}}{h_{20}} \frac{d\Phi_2}{d\hat{q}_2} \oint_\Gamma \mathbf{u}_B \cdot \mathbf{s}\,ds, \qquad (6.46)$$

where $\mathbf{j}$ is the meridional unit vector. Thus

$$\frac{d\Phi_2}{d\hat{q}_2} = \frac{h_{20}}{\hat{F}}, \qquad (6.47)$$

and because $\hat{q}_2 = \beta_2 r_1$ on the outermost closed contour $\Gamma_0$,

$$\Phi_2(\hat{q}_2) = \frac{h_{20}}{\hat{F}}(\hat{q}_2 - \beta_2 r_1) = \frac{\beta}{\hat{F}}(y - r_1) + \phi_B \qquad (6.48)$$

in the region $A_0$ enclosed by the closed geostrophic contours. In this region, the potential vorticity $q_2$ is

$$q_2(x, y) = \hat{q}_2(x, y) - \frac{\hat{F} + F_2}{h_{20}}\phi_2(x, y) = \beta_2 r_1 - \frac{F_2}{h_{20}}\phi_2(x, y), \qquad (6.49)$$

where $\phi_2(x, y) = \Phi_2[\hat{q}_2(x, y)]$. Thus, for dissipation in the form of the vorticity drag (6.42), the subsurface layer 2 is again set in motion in the region of closed geostrophic contours, but potential vorticity $q_2$ in this region is not homogenized. Outside the region of closed contours, the solutions for $q_2$ and $\phi_2$ are again (6.40) and (6.41), respectively.

## 6.3 Potential Vorticity Homogenization: Planetary Geostrophic Theory

Some of the arguments of the preceding section can be generalized to the planetary geostrophic case of large thickness fluctuations and large meridional extents, removing the restrictions (6.9) and (6.10). The additional nonlinearity, however, makes explicit solution of the resulting equations more difficult.

In the planetary geostrophic case, a stream function for the horizontal velocities does not exist, but a transport stream function $\psi_j$ can still be found for steady flow in subsurface layer $j$, according to (4.76). Thus the previous area integral over the region bounded by a closed streamline can be replaced by a volume integral over the region $A_j$ bounded by any closed transport streamline $\Gamma_j$ on which $\psi_j = $ constant and by the layer interfaces above and below, with the result that

$$
\int_{A_j} \mathbf{u}_j \cdot \nabla_h Q_j \, dx \, dy \, dz = \int_{A_j} h_j \mathbf{u}_j \cdot \nabla_h Q_j \, dx \, dy
$$

$$
= \int_{A_j} \nabla_h \cdot (h_j \mathbf{u}_j Q_j) \, dx \, dy
$$

$$
= \oint_{\Gamma_j} h_j \mathbf{u}_j Q_j \cdot \mathbf{n} \, ds = 0 \tag{6.50}
$$

because the velocity on the streamline is perpendicular to the normal $\mathbf{n}$ to the streamline.

Thus, if a small dissipative term $\mathcal{D}_j$ is again added to the steady potential vorticity balance so that

$$
\mathbf{u}_j \cdot \nabla_h Q_j = \mathcal{D}_j, \tag{6.51}
$$

it follows both that

$$
Q_j = \hat{Q}_j(\psi_j) + \mathcal{O}(\mathcal{D}_j) \tag{6.52}
$$

and, by (6.50), that

$$
\int_{A_j} \mathcal{D}_j \, dx \, dy \, dz = \int_{A_j} h_j \mathcal{D}_j \, dx \, dy = 0. \tag{6.53}
$$

Suppose now that the dissipative term takes the specific form

$$
\mathcal{D}_j = \frac{K_Q}{h_j} \nabla_h^2 Q_j, \tag{6.54}
$$

where $K_Q$ is a constant, so that the vertically integrated down-gradient flux of potential vorticity $-K_Q \nabla Q_j$ is independent of the thickness of the layer. The quantity $K_Q/h_j$ is an eddy diffusion coefficient; the requirement that $\mathcal{D}_j$ be small translates into the condition that $K_Q/h_j \ll UL$, where $U$ and $L$ are horizontal velocity and length scales for the large-scale flow. With $\mathcal{D}_j$ defined by (6.54), the integral of the dissipative term gives

$$
\int_{A_j} K_Q \nabla_h^2 Q_j \, dx \, dy = K_Q \int_{A_j} \nabla_h^2 Q_j \, dx \, dy
$$

$$
= K_Q \oint_{\Gamma_j} \nabla_h Q_j \cdot \mathbf{n} \, ds
$$

$$
= K_Q \frac{d\hat{Q}_j}{d\psi_j} \oint_{\Gamma_j} \nabla_h \psi_j \cdot \mathbf{n} \, ds \tag{6.55}
$$

to first order in $K_Q$, where the last equality follows because $\psi_j$ is constant on $\Gamma_j$. Because the gradient of the stream function $\psi_j$ is parallel to the normal $\mathbf{n}$ to the streamline, (6.50) and (6.55) imply, again to first order in $K_Q$, that

$$\frac{d\hat{Q}_j}{d\psi_j} = 0 \quad \text{on} \quad \Gamma_j. \tag{6.56}$$

It follows that the potential vorticity $Q_j$ must be constant throughout the region $A_{j0}$ bounded by the outermost closed streamline $\Gamma_{j0}$. Thus, with the assumption (6.54) for the form of the dissipative term in the potential vorticity equation, homogenization of potential vorticity inside a region of closed streamlines is implied also in the planetary geostrophic case.

An example of a subsurface region in which the homogenization argument might apply is the unventilated, western pool region of layer 2 in the two-layer ventilated thermocline model of subtropical gyre circulation discussed in the preceding chapter. That region is bounded to the east by the trajectory $x = x_P(y)$, which is a contour of constant layer 2 potential vorticity. It is natural to consider the possibility, motivated by the preceding argument, that the layer 2 potential vorticity in this western pool region should be uniform and equal to the potential vorticity $f_2/H_{2W}$ on $x = x_P(y)$. This deduction, however, rests on the assumptions that the trajectory $x = x_P(y)$ is closed in the western boundary current and that the potential vorticity dissipation along the western boundary segment of the trajectory takes the same form (6.54). The western pool structure that results from these assumptions is discussed in the preceding chapter.

A second, related example involves flow in the layers beneath the deepest wind-driven layers of the ventilated thermocline model. For example, in the two-layer model, consider the structure of potential vorticity contours in the motionless layer 3 near the subtropical-subpolar gyre boundary $y = y_b$, where $W_E = 0$, and north of the outcrop $y = y_2$ so that $h_1 = 0$. Suppose that layer 3 has a constant, finite thickness $h_3(x_E, y) = h_{3E}$ at the eastern boundary $x = x_E$, where the layer 2 thickness $h_2(x_E, y) = h_E$. Then the layer 3 potential vorticity $Q_3$ is

$$Q_{3E}(x_E, y) = \frac{f}{h_{3E}} \tag{6.57}$$

at the eastern boundary, while

$$Q_3(x, y) = \frac{f}{h_3} = \frac{f}{H_3 - H_2} = \frac{f}{h_E + h_{3E} - H_2(x, y)} \tag{6.58}$$

in the interior, where

$$H_2(x, y) = h_2(x, y) = \left[ h_E^2 + D_2^2(x, y) \right]^{1/2}. \tag{6.59}$$

The contour $x = x_Q(y)$ of constant layer 3 potential vorticity that intersects the eastern boundary at $y = y_Q$ is then defined implicitly by

$$Q_3[x_Q(y), y] = Q_{3E}(y_Q) = \frac{f_Q}{h_{3E}}, \tag{6.60}$$

where $f_Q = f(y_Q)$. If the imposed Ekman pumping depends only on $y$, so that $W_E = W_0(y)$, an explicit solution may be obtained for $x_Q(y)$:

$$x_Q = x_E + \frac{\beta \gamma_2}{2 f^2 W_0(y)} \left[ 2 \left( 1 - \frac{f}{f_Q} \right) h_E h_{3E} + \left( 1 - \frac{f}{f_Q} \right)^2 h_{3E}^2 \right]. \tag{6.61}$$

Provided that $W_0(y) \neq 0$, $x_Q(y) \to x_E$ as $y \to y_Q$, and the contour intersects the eastern boundary as anticipated. Suppose, however, that $y_Q = y_b$ so that $W_0(y_Q) = 0$. Then, with $f_b = f(y_b)$,

$$x_Q(y; y_Q = y_b) = x_{Qb} \to x_E - \frac{\beta^2 \gamma_2}{f_b^3 W_0'(y_b)} h_E h_{3E} \quad \text{as} \quad y \to y_b, \tag{6.62}$$

and because $W_0'(y_b) > 0$ in general, the contour $x = x_{Qb}(y)$ instead reaches the latitude $y = y_b$ at a finite distance westward of the eastern boundary. Consequently, the potential vorticity contours west of $x = x_{Qb}(y)$ are isolated from the eastern boundary and will instead emanate from and return to the western boundary. In this way, this region $x < x_{Qb}(y)$ in the otherwise motionless layer 3 is analogous to the western pool region in layer 2, and by a similar argument, the potential vorticity in this recirculation region of layer 3 may be presumed to homogenize.

A physical interpretation of (6.62) may be given that is similar to the interpretation (6.30) of the condition (6.29) for the development of closed potential vorticity contours in layer 2 in the quasi-geostrophic example discussed in the preceding section. Define a vertically averaged zonal Sverdrup transport velocity at $y = y_Q = y_b$ for layers 2 and 3 as

$$\begin{aligned}
u_S(x, y_b) &= \frac{h_2 u_2}{h_E + h_{3E}} \\
&= \frac{1}{h_E + h_{3E}} \left[ -\frac{\gamma_2}{2 f_b} \frac{\partial}{\partial y} D_2^2(x, y) \right]_{y=y_b} \\
&= -\frac{1}{h_E + h_{3E}} \frac{f_b}{\beta} W_0'(y_b)(x - x_E). \tag{6.63}
\end{aligned}$$

Then, at $x = x_Q$, with (6.62),

$$u_S(x_Q, y_b) = \frac{\beta \gamma_2}{f_b^2} \frac{h_E h_{3E}}{h_E + h_{3E}} = -c_R, \tag{6.64}$$

where $c_R$ is the phase speed of the first-mode long baroclinic planetary wave on layers 2 and 3 at $y = y_b$. Thus, as in the quasi-geostrophic example, the condition for formation of the region of closed potential vorticity contours is that the Sverdrup flow be equal and opposite to the phase speed of the long baroclinic Rossby wave. The point at which this occurs is also the point at which the meridional gradient of potential vorticity in the lower layer vanishes because

$$\frac{\partial Q_3}{\partial y}(x, y) = \frac{\partial}{\partial y}\frac{f}{h_3}$$

$$= \frac{1}{h_3}\left(\beta - \frac{f}{h_3}\frac{\partial h_3}{\partial y}\right)$$

$$= \frac{\beta}{h_3}\left(1 - \frac{f^2 u_2}{\beta\gamma_2 h_3}\right), \tag{6.65}$$

from which, with (6.63) and (6.64), it follows that

$$\frac{\partial Q_3}{\partial y}(x_Q, y_b) = 0. \tag{6.66}$$

## 6.4 Notes

The homogenization theory in the absence of planetary rotation is originally from Batchelor (1956) and had been noted earlier by Prandtl (1905). The potential vorticity homogenization theory for ocean circulation was developed under quasi-geostrophic assumptions by Rhines and Young (1982a, 1982b) and discussed for planetary geostrophic flows by Pedlosky and Young (1983).

# 7

# Circumpolar Flow

## 7.1 The Geostrophic Constraint

Suppose now that the basin geometry is altered from the simple, closed domain considered in the preceding chapters to contain a representation of the circumpolar connection that exists around Antarctica in the Southern Ocean. The simplest such representation is a periodic, reentrant zonal channel in the southern portion of an otherwise closed southern hemisphere domain, with width comparable to that of the opening between South America and Antarctica at Drake Passage (Figure 7.1). Even if the channel is partly blocked by a sill along the domain boundary at some depth above the basin bottom, this is a highly idealized geometry: in the Southern Ocean, much of the latitude-depth cross section of the Drake Passage opening is obscured at other longitudes by islands or shallow bottom topography, which can be anticipated to have a strong influence on both the circumpolar and the meridional flow. Nonetheless, this geometry allows the exploration of several fundamental elements of the effect of the circumpolar connection on the large-scale flow.

Let the periodic zonal channel representing the circumpolar connection cover the latitude range $[y_-, y_+]$ and the depth range $[-H_s, 0]$, where $H_s \leq H_0$ is the depth of the sill at $x = x_E$ or $x = x_W$. In the channel opening, periodic boundary conditions apply, and the points at $x = x_E$ and $x = x_W$ are identified, so that $\phi(x = x_E) = \phi(x = x_W)$ for any dynamical variable $\phi$. At any latitude $y = y_\Gamma$ in the periodic zonal channel (i.e., at any $y_\Gamma$ such that $y_- < y_\Gamma < y_+$) and depth $z > -H_s$ above the sill, the geostrophic zonal momentum balance (4.7) can be integrated around a closed, circumpolar contour $\Gamma$ of constant latitude $y = y_\Gamma$, with the result that

$$\oint_\Gamma f v(x, y_\Gamma, z)\, dx = f_\Gamma \oint_\Gamma v\, dx = \oint_\Gamma \frac{\partial p'}{\partial x}\, dx \equiv 0, \qquad (7.1)$$

where $f_\Gamma = f(y_\Gamma)$. This implies that the zonally integrated geostrophic meridional transport $V_\Gamma(y_\Gamma, z)$ at any level $z > -H_s$ in the circumpolar gap $[y_-, y_+]$ must

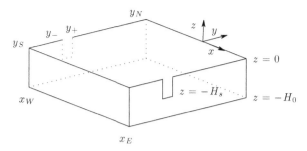

Figure 7.1. Domain geometry for the rectangular, single-hemisphere basin with circumpolar connection. The lateral boundaries are vertical, rigid walls at $x = x_E$, $x = x_W$, $y = y_S$, and $y = y_N$, except in the gap region $y_- < y < y_+$, $-H_s < z < 0$, where zonally periodic boundary conditions are imposed at $x = x_E$ and $x = x_W$, allowing circumpolar flow. The rigid, flat top and bottom are at $z = 0$ and $z = -H_0$, respectively.

vanish:

$$V_\Gamma(y_\Gamma, z) = \oint_\Gamma v(x, y_\Gamma, z)\, dx = 0, \quad y_\Gamma \in [y_-, y_+]. \tag{7.2}$$

Thus, as a result of the geostrophic constraint (7.1), there can be no zonally integrated meridional geostrophic transport above the sill at the channel latitudes.

The geostrophic constraint applies to the total meridional geostrophic volume transport but not to the geostrophic heat transport. Consider, for example, a three-layer model in which the lower layer is at rest and the upper two layers are located above the sill depth $z = -H_s$ so that transport integrals may be computed around the closed contour $\Gamma$. Then the zonally integrated meridional geostrophic mass transports in each layer will be equal and opposite:

$$\oint_\Gamma h_2 v_2\, dx = \oint_\Gamma (H_2 - H_1)\frac{\gamma_2}{f_\Gamma}\frac{\partial H_2}{\partial x}\, dx = -\frac{\gamma_2}{f_\Gamma}\oint_\Gamma H_1\frac{\partial H_2}{\partial x}\, dx \tag{7.3}$$

$$\oint_\Gamma h_1 v_1\, dx = \oint_\Gamma H_1\left(\frac{\gamma_2}{f_\Gamma}\frac{\partial H_2}{\partial x} + \frac{\gamma_1}{f_\Gamma}\frac{\partial H_1}{\partial x}\right) dx = \frac{\gamma_2}{f_\Gamma}\oint_\Gamma H_1\frac{\partial H_2}{\partial x}\, dx. \tag{7.4}$$

Thus the total vertically integrated transport will vanish even if the individual layer components do not. The opposite meridional volume transports in the two layers of different temperatures imply a net meridional heat transport equal to

$$\mathcal{H} = \rho_0 C_p\left(T_1 \oint_\Gamma h_1 v_1\, dx + T_2 \oint_\Gamma h_2 v_2\, dx\right)$$

$$= \rho_0 C_p(T_1 - T_2)\frac{\gamma_2}{f_\Gamma}\oint_\Gamma H_1\frac{\partial H_2}{\partial x}\, dx. \tag{7.5}$$

The latter integral may be recast in terms of the pressure force from layer 2 on the layer 1 interface:

$$\gamma_2 \oint_\Gamma H_1 \frac{\partial H_2}{\partial x} \, dx = -\gamma_2 \oint_\Gamma H_2 \frac{\partial H_1}{\partial x} \, dx = -\frac{1}{\rho_0} \oint_\Gamma p_2 \frac{\partial H_1}{\partial x} \, dx. \qquad (7.6)$$

This integral of the pressure force on the interior layer interface is sometimes referred to as the internal form stress. By (7.3) and (7.6), this pressure force is balanced in the layer-integrated momentum balance by the Coriolis force associated with the net meridional mass flux in layer 2:

$$\rho_0 f_\Gamma \oint_\Gamma h_2 v_2 \, dx = \oint_\Gamma h_2 \frac{\partial p_2}{\partial x} \, dx = \oint_\Gamma p_2 \frac{\partial H_1}{\partial x} \, dx. \qquad (7.7)$$

Note that there must be zonal variations in both $H_1$ and $H_2$ in order that the integrated meridional geostrophic heat flux $\mathcal{H}$ not vanish identically.

## 7.2 Depth-Integrated Flow with Linear Frictional Closure

The effect that the geostrophic constraint (7.2) has on the large-scale circulation can be illustrated by considering the solution, in a basin with a circumpolar channel, of the model (3.38) of depth-integrated large-scale flow with linear friction that was previously solved in the simple closed basin. The solution for the depth-integrated flow with this simple frictional closure demonstrates, through explicit example, that the circumpolar flow depends on dynamical balances that are essentially different from the Sverdrup transport balance, even in the limit in which the friction parameter in the example is made vanishingly small.

For simplicity, let the circumpolar gap extend to the southern boundary of the domain and have no sill so that $y_- = y_S$ and $H_s = H_0$. For given wind forcing, the flow may be determined by solving the vorticity equation (3.40) for the Sverdrup stream function $\Psi$ (3.18). However, the circumpolar gap introduces a new component to the problem: on each contiguous segment of the boundary, the no-normal-flow condition requires that $\Psi$ be constant, but now there are two such segments, on which $\Psi$ may take different constant values.

The additional equation, supplementing the vorticity equation (3.40), that is required to close the problem can be obtained by integrating the depth-integrated zonal momentum equation in (3.38) over the area $A_{ch} = [x_W, x_E] \times [y_S, y_+]$ of the channel:

$$0 = \int_{A_{ch}} \left[ -f \frac{\partial \Psi}{\partial x} + \frac{\partial \bar{P}}{\partial x} - \tau_w^x - r \frac{\partial \Psi}{\partial y} \right] dx \, dy. \qquad (7.8)$$

Because the zonal integral of the first two terms along any closed, fixed-latitude contour vanishes identically, this equation reduces to a balance between forcing and friction. This balance determines the difference between the constant value $\Psi_S = \Psi(x, y_S)$ of the transport stream function on the southern boundary and its average value $\Psi_b$ along the northern edge $y = y_+$ of the channel:

$$\Psi_b - \Psi_S = -\frac{1}{r(x_E - x_W)} \int_{y_S}^{y_+} \oint \tau_w^x \, dx \, dy, \tag{7.9}$$

where

$$\Psi_b = \frac{1}{x_E - x_W} \oint \Psi(x, y_+) \, dx. \tag{7.10}$$

For small $r$, $\Psi_b$ will be essentially determined by the Sverdrup solution for $y > y_+$, with a zonal flow in the channel region $y < y_+$ that is determined by the balance (7.9) of zonal wind stress and friction (Figure 7.2). Thus the mean zonal transport through the channel in this model will be approximately proportional to the mean zonal wind stress and inversely proportional to the friction coefficient $r$. In contrast to the closed-basin solution, in which the interior solution is determined by the Sverdrup balance, the role of friction in this case is therefore not confined to boundary layers as the amplitude of the interior response to forcing depends directly on the value of the friction parameter $r$.

Explicit solutions of this problem for small $r$ can be obtained using boundary layer theory, as in the case of the closed basin. The solution is essentially the same as that for the closed basin in the region north of the channel, $y > y_+$. Along the northern edge of the channel $y = y_+$, there is a parabolic boundary layer with westward-increasing meridional width $[r(x - x_E)]^{1/2}$; a similar boundary layer is found in the closed basin when the wind forcing does not vanish along a rigid zonal boundary. At the point $(x_E, y_+) = (x_W, y_+)$, the southern tip of the northern boundary that extends southward to the northern edge of the channel, there is a new type of boundary layer with infinite velocity at the tip but finite transport. With, for example, $\tau_w^x = \tau_0$ a constant in $y_S < y < y_+$, the interior solution in the channel has uniform zonal flow, $\Psi(x, y) = \Psi_S - \tau_0(y - y_S)/r$. For arbitrarily small $r$, this interior solution illustrates that the zonal transport in the channel will become arbitrarily large because $\Psi_b - \Psi_S \to \infty$ as $r \to 0$.

Solutions of this model for reasonable values of the friction parameter $r$ yield circumpolar currents that are much larger than observed. Thus this model proves inadequate even as a first-order representation of the large-scale dynamics of the Antarctic Circumpolar Current. Nonetheless, it provides a useful illustration of the dramatic effect of the circumpolar connection on the large-scale circulation and an explicit example in which it can be seen directly that the vertically integrated circumpolar current transport is not determined by the Sverdrup transport balance.

## 7.3 Circumpolar Flow over a Topographic Sill

Because the Sverdrup stream function (3.18) for the total depth-integrated flow may still be defined when the bottom is not flat, and because the southern boundary of the

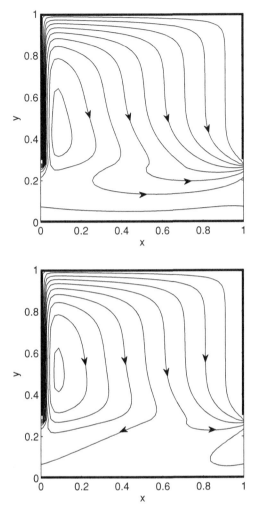

Figure 7.2. Stream function contours $\Psi = \text{const}$ for dimensionless numerical solutions of the model equations (3.40) and (7.9) for the depth-integrated wind-driven gyre circulation with a circumpolar connection in the southern part of a southern hemisphere basin for $\delta_S = 0.02$. The rigid lateral boundaries are indicated by thick solid lines. The circumpolar connection, with zonally periodic boundary conditions, occupies $0 < y < 0.3$ at $x = 0$ and $x = 1$. (top) Solution with eastward zonal wind forcing at the circumpolar connection latitudes. Lower panel: Solution with no eastward zonal wind forcing at the circumpolar connection latitudes. In both cases, the contour interval is 0.1, and $\Psi = 0$ along the contiguous northern boundary and increases away from the boundary. (Redrawn from Samelson [1999]. Original figure © American Meteorological Society. Used with permission.)

basis is rigid, the integral of the total meridional transport across the constant-latitude circumpolar contour $\Gamma$ can be computed as

$$\oint_\Gamma \int_{-H(x,y)}^0 (v + v_E)\, dz\, dx = \oint_\Gamma V_s\, dx = \oint_\Gamma \frac{\partial \Psi}{\partial x}\, dx = 0. \qquad (7.11)$$

Thus, in addition to the constraint (7.1) on the geostrophic meridional transport, the zonal and vertical integral of the total meridional transport across the closed zonal contour at a fixed latitude $y_\Gamma$ must also vanish.

Suppose now that $H_s < H_0$ so that the circumpolar gap has a topographic sill. The geostrophic constraint does not hold below the sill depth because the zonal integral of pressure at fixed depth is blocked by the topography. However, by (7.11), the integrated Coriolis force at a fixed latitude $y = y_\Gamma$ in the circumpolar gap vanishes identically from the vertically and zonally integrated zonal momentum balance so that

$$0 = \oint_\Gamma \int_{-H(x,y)}^0 \left[ \rho_0 f(v + v_E) - \frac{\partial p'}{\partial x} - \frac{\partial \tau_E^x}{\partial z} \right] dz\, dx$$

$$= -\oint_\Gamma \int_{-H(x,y)}^0 \frac{\partial p'}{\partial x}\, dz\, dx + \oint_\Gamma \tau_w^x(x, y)\, dx, \qquad (7.12)$$

where it has been assumed that the turbulent stress $\tau_E^x$ vanishes beneath the Ekman depth $\delta_E$ and that $\delta_E \ll H_s$. Because, by (7.1),

$$\oint_\Gamma \int_{-H_s}^0 \frac{\partial p'}{\partial x}(x, y, z)\, dz\, dx = \int_{-H_s}^0 \oint_\Gamma \frac{\partial p'}{\partial x}(x, y, z)\, dx\, dz = 0, \qquad (7.13)$$

it follows that

$$\oint_\Gamma \int_{-H(x,y)}^{-H_s} \frac{\partial p'}{\partial x}(x, y, z)\, dz\, dx - \oint_\Gamma \tau_w^x(x, y)\, dx = 0. \qquad (7.14)$$

Thus the large-scale dynamics require that the vertically and zonally integrated zonal pressure gradient force below the sill depth balance the zonally integrated zonal wind stress at the surface.

The pressure integral in (7.14) may be written as the zonally integrated zonal pressure force on the bottom topography:

$$\oint_\Gamma \int_{-H(x,y)}^{-H_s} \frac{\partial p'}{\partial x}(x, y, z)\, dz\, dx = -\oint_\Gamma p'(x, y, -H) \frac{\partial H}{\partial x}\, dx. \qquad (7.15)$$

Equation (7.14) then shows that the zonally integrated zonal wind stress at the surface must be balanced by this pressure force:

$$\oint_\Gamma p'(x, y, -H) \frac{\partial H}{\partial x}\, dx + \oint_\Gamma \tau_w^x(x, y)\, dx = 0. \qquad (7.16)$$

The integral of the pressure force on the bottom topography is sometimes referred to as *topographic form stress* or *form drag*.

An alternate form of (7.14) may be obtained by dividing both sides of the equation by $\rho_0 f_\Gamma$ and substituting the geostrophic balance (4.7) and Ekman transport (3.21) relations. This yields

$$\oint_\Gamma \int_{-H(x,y)}^{-H_s} v(x, y_\Gamma, z)\, dz\, dx + \oint_\Gamma V_E(x, y_\Gamma)\, dx = 0, \qquad (7.17)$$

where $v$ is the meridional geostrophic velocity. Note that the geostrophic transport integral is only over depths beneath the sill depth because, by (7.2), the integrated geostrophic transport above the sill depth vanishes identically. Equation (7.17) has the form of a mass balance: the sum of the integrated meridional geostrophic and Ekman transports across the fixed latitude $y = y_\Gamma$ must vanish. This mass balance (7.17) can be obtained directly from the integral (7.11) of the Sverdrup transport. Note also that because the zonal integral of the Sverdrup stream function along $\Gamma$ vanishes identically, the zonal integral of the topographic Sverdrup transport balance (3.54) is

$$0 = \oint_\Gamma \left[ \rho_0 \beta \frac{\partial \Psi}{\partial x} - J(p_b, H) - \left( \frac{\partial \tau_w^y}{\partial x} - \frac{\partial \tau_w^x}{\partial y} \right) \right] dx$$

$$= \oint_\Gamma \left[ \frac{\partial}{\partial y} \left( p_b \frac{\partial H}{\partial x} \right) + \frac{\partial \tau_w^x}{\partial y} \right] dx, \qquad (7.18)$$

which is just the meridional derivative of the integrated zonal momentum balance (7.16).

Two essential conclusions follow from these considerations. First, the vertically integrated large-scale zonal momentum balance relates the meridional geostrophic flow to the zonal wind stress but does not provide any direct relation between the zonal wind stress and the zonal transport of the circumpolar flow nor any other way to determine that zonal transport. Second, although the Sverdrup vorticity relation (3.8) is presumed to hold in the circumpolar flow regime, the Sverdrup transport balance also does not determine the circumpolar zonal transport; instead, like the momentum balance, it reduces to a relation between wind stress, bottom pressure, and bottom topography.

In the Southern Ocean, closed, circumpolar contours at fixed latitude and depth can be found to depths of roughly 1500 m in narrow meridional intervals between 59° S and 60° S. Thus the constraint (7.2) on meridional geostrophic transport across the model circumpolar gap and the integrated zonal momentum balance correctly represent fundamental elements of the large-scale circulation dynamics of the Southern Ocean. It follows that additional dynamical considerations, beyond the large-scale dynamical balances that lead to (7.16) or (7.17), are necessary to develop even the simplest large-scale theory for the existence and amplitude of the circumpolar flow that is observed in the Southern Ocean: unlike the case of Sverdrup theory for the wind-driven gyres, knowledge of the wind stress field alone is not sufficient to provide a theoretical estimate of the strength and structure of the depth-integrated large-scale flow.

## 7.4 A Baroclinic Model of Circumpolar Flow

A proper understanding of the fundamental dynamics of the circumpolar current that may develop in a basin with a circumpolar connection requires consideration of thermal as well as wind forcing. Special forms of wind and thermal forcing can be chosen that simplify the analysis and focus attention on the role of the circumpolar connection and or the dependence of the resulting circumpolar current on geometric and forcing parameters. The essential element of the wind forcing consists of a meridional band of westerlies centered at the gap latitudes. With this imposed eastward stress field, the resulting surface Ekman transport is directed northward across the gap because the Coriolis parameter $f$ is negative in the southern hemisphere. The essential element of the surface thermal forcing consists of a fixed meridional temperature gradient across the gap, with warmer, less dense fluid on the northern (equatorward) side of the gap and cooler, denser fluid on the southern (poleward) side. This combination of surface forcing and basin geometry is sufficient to produce, through the large-scale dynamics, a thermal circumpolar current that flows through the circumpolar connection in the rectangular basin.

These ideas can be illustrated with a simple quantitative dynamical model that makes the dependence on the thermal and geometric parameters explicit. Consider the thermodynamic equation in its adiabatic large-scale form (4.10) and for steady flow. Let the ocean-surface temperature $T_s$ be specified as

$$T_s = \begin{cases} T_-, & y_S \leq y < y_-, \\ T_- + G_T (y - y_-), & y_- \leq y < y_+, \\ T_+, & y_+ \leq y < y_N, \end{cases} \tag{7.19}$$

where

$$G_T = \frac{T_+ - T_-}{y_+ - y_-} \tag{7.20}$$

is the constant meridional temperature gradient across the gap and where $T_+ > T_-$ (Figure 7.3). Let the wind stress be purely zonal with meridional dependence such that the Ekman vertical velocity $W_E = dV_E/dy$ is

$$W_E = \begin{cases} W_- \sin\left[\pi \left(\frac{y - y_S}{y_- - y_S}\right)\right], & y_S \leq y < y_-, \\ 0, & y_- \leq y < y_+, \\ W_+ \sin\left[\pi \left(\frac{y - y_+}{y_N - y_+}\right)\right], & y_+ \leq y < y_N, \end{cases} \tag{7.21}$$

where $W_- > 0$ and $W_+ < 0$ are constants that satisfy $W_-(y_- - y_S) + W_+(y_N - y_+) = 0$ so that the Ekman circulation conserves total volume. Thus there is upward Ekman pumping south of the gap, no Ekman pumping within the gap, and downward Ekman pumping north of the gap. Because of the surface thermal boundary condition

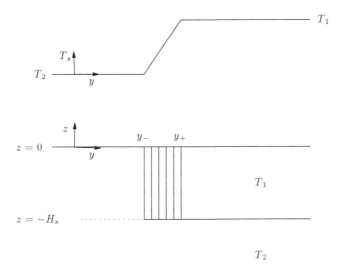

Figure 7.3. (top) Surface thermal conditions $T_s(y)$ vs. latitude $y$ and (bottom) meridional cross section of ocean isotherms vs. latitude $y$ and depth $z$ for the analytical
model (7.19)–(7.28) of a thermal circumpolar current. The circumpolar connection
occupies the region $y_- < y < y_+$, $-H_s < z < 0$. The surface temperature takes the
uniform values $T_1$ and $T_2$ north and south of the gap, respectively, with a constant
gradient across the gap. Everywhere south of the gap or below $z = -H_s$, the ocean
temperature is equal to the cold southern surface value $T_2$. In the region north of the
gap and above $z = -H_s$, the ocean temperature is equal to the warm surface value
$T_1$. The ocean temperature at each point $(y, z)$ in the gap is equal to the corresponding
surface temperature $T_s(y)$ in the gap.

(7.19), the latter represents a continuous source of warm ($T = T_+$) fluid entering the
subtropical gyre north of the gap ($y > y_+$).

With the surface temperature $T_s$ given by (7.19), it follows that the fluid south of
the gap ($y < y_-$) must have $T = T_-$ everywhere because $T_s = T_-$, and any warmer
fluid will be brought to the surface by convective adjustment and immediately cooled.
Thus the motion in the southern wind-forced region ($y_S < y < y_-$) is barotropic. A
zonal integral of the geostrophic Sverdrup balance $\beta V_G = f W_E$ gives the interior
surface geostrophic pressure $p_s(x, y)$ in $y_S \le y < y_-$:

$$p_s(x, y) = -\frac{\rho_0 f^2 W_-}{\beta}(x_E - x) \sin\left[\pi\left(\frac{y - y_S}{y_- - y_S}\right)\right]. \qquad (7.22)$$

Because $p_s(x, y_S) = p_s(x, y_-) = p_s(x_E, y) = 0$ and $p_s$ decreases westward, contours
of $p_s$ will have the standard, roughly semicircular gyre structure, with open contours
adjacent to the assumed western boundary layer. The corresponding geostrophic
motion around the gyre will be cyclonic (clockwise in the southern hemisphere),
with eastward motion adjacent to the gap, westward motion adjacent to the southern
boundary, and southward motion everywhere.

North of the gap ($y > y_+$), it is natural to make the traditional assumption that the geostrophic no-normal-flow condition may be applied directly to the interior flow along the eastern boundary so that isotherms along the eastern boundary must be flat. Because the sidewalls extend continuously around the basin up to the sill depth $z = -H_s$, it follows that the fluid beneath the sill depth ($z < -H_s$) must have $T = T_-$ along the entire eastern boundary. In contrast, the fluid pumped downward from the surface layer north of the gap ($y > y_+$) has uniform temperature $T = T_+$, and the geostrophic constraint prevents any southward return flow of this warm fluid across the gap above the sill depth. Along the eastern boundary, then, the warm ($T = T_+$) fluid will extend downward all the way to the sill depth $z = -H_s$, where it first encounters opposition from cold ($T = T_-$) fluid able to flow northward geostrophically.

Consequently, the fluid north of the gap divides into two layers: a lower layer that has $T = T_-$ and is at rest and an upper layer of thickness $h(x, y)$ that has $T = T_+$ and is driven by Ekman pumping, with eastern boundary condition $h(x_E, y) = h_E = H_s$. The solution for the moving, upper layer north of the gap will then be that for the one-layer wind-driven model with $h_E = H_s$:

$$h(x, y) = \left[ H_s^2 + D_1^2(x, y) \right]^{1/2}, \tag{7.23}$$

where, for $y_+ < y < y_N$,

$$D_1^2(x, y) = -\frac{2 f^2}{\beta \gamma_1} (x_E - x) W_+ \sin \left[ \pi \left( \frac{y - y_+}{y_N - y_+} \right) \right], \tag{7.24}$$

where now $\gamma_1 = g \alpha_T (T_+ - T_-)$. The surface pressure $p_s(x, y) = \gamma_1 h(x, y)$, and because $h(x, y_+) = h(x, y_N) = h(x_E, y) = H_s$ and $h$ increases westward, contours of $h$ and $p_s$ will again have the standard subtropical gyre structure, with open contours adjacent to the western boundary. The corresponding geostrophic motion around this subtropical gyre will be anticyclonic (counterclockwise in the southern hemisphere), with eastward motion adjacent to the gap, westward motion adjacent to the northern boundary, and northward motion everywhere.

When the warm fluid north of the gap reaches the northern edge of the gap just below the sill depth, it will flow southward across the gap in a deep western boundary current to compensate the northward surface Ekman transport across the gap and close the meridional circulation cell. However, as it does so, it will flow beneath colder surface fluid because of the imposed surface thermal gradient across the gap. Convective adjustment must be invoked to remove this gravitational instability. The effect of this adjustment is that the water column at the sill will assume the local surface temperature throughout the depth range from the surface $z = 0$ to the sill depth $z = -H_s$ (Figure 7.3):

$$T(x \approx x_W, y, z) = T_s(y) \quad \text{for } y_- < y < y_+, \quad -H_s \leq z \leq 0. \tag{7.25}$$

By the thermal wind balance, a zonal geostrophic flow will be associated with this temperature field. Because a purely zonal geostrophic flow with vertically uniform temperature is an exact solution of the planetary geostrophic equations (with $\kappa_V = 0$), this temperature field may be extended zonally across the basin so that

$$T(x, y, z) = T_s(y) \quad \text{for} \quad y_- < y < y_+, \quad -H_s \le z \le 0. \tag{7.26}$$

Away from the deep boundary current, the deep, cold $(T = T_-)$ interior fluid beneath the sill is stagnant, so the pressure at the sill depth $z = -H_s$ is uniform. The hydrostatic relation may therefore be integrated vertically from the sill depth to construct the pressure field $p'$ in the gap, giving a purely zonal geostrophic velocity field:

$$p'(y, z) = \rho_0 g \alpha_T T_s(y)(z + H_s) \tag{7.27}$$

$$u(y, z) = -\frac{g \alpha_T G_T}{f}(z + H_s) \tag{7.28}$$

for $y_- < y < y_+$, $H_s \le z \le 0$. Because $G_T > 0$ and $f < 0$, the flow described by (7.28) is eastward $(u > 0)$. The total integrated zonal transport $U_c$ of this circumpolar current is

$$U_c = \int_{y_-}^{y_+} \int_{-H_s}^{0} u \, dy \, dz = g \alpha_T G_T \frac{H_s^2}{2\beta} \ln \frac{f_-}{f_+} \approx -g \alpha_T (T_+ - T_-) \frac{H_s^2}{2 f_+}, \tag{7.29}$$

where $f_- = f(y_-)$ and $f_+ = f(y_+)$. The previous solutions north and south of the gap show that the gyre flow adjacent to the gap is also zonal and eastward but is much weaker than the zonal jet in the gap. The meridional circulation is assumed to be closed by a deep western boundary current at the sill, just beneath the sill depth.

It follows from (7.28) and (7.29) that in this model, the zonal velocity $u$ and total integrated zonal transport $U_c$ of the circumpolar current depend only on thermal and geometric parameters. This perhaps surprising conclusion is consistent with the preceding discussion of the geostrophic constraint, in which it was shown that the wind forcing, through the Sverdrup relation or the zonal momentum balance, does not determine the transport of the circumpolar current in the large-scale dynamical context. Instead, the current in this model is a thermal geostrophic current that owes its existence entirely to the meridional gradient of ocean temperature at the gap latitudes. The subsurface meridional temperature gradient arises as a result of surface warming and subsequent downward Ekman pumping of fluid transported northward across the gap in the wind-driven surface Ekman layer. Thus the existence of westerly wind forcing in the gap latitudes is essential for the formation of the current even in this model: without the wind forcing, the warm fluid north of the gap could be confined to an infinitesimally thin surface layer, resulting in no interior zonal geostrophic flow. Provided there is sufficient westerly wind forcing to drive the meridional surface

Ekman flow, however, the resulting circumpolar geostrophic current is independent of the strength of the wind in this large-scale geostrophic model. Instead, the amplitude of the current depends on the meridional difference in surface temperature across the gap and on the width and depth of the gap.

Although it contains only the simplest abstract elements of Southern Ocean geometry, forcing, and dynamics, this construction is perhaps the simplest physically based model for the existence of the Antarctic Circumpolar Current. For dimensional scales $T_+ - T_- = 5\,\mathrm{K}$, $y_+ - y_- = 500\,\mathrm{km}$, and $H_s = 2000\,\mathrm{m}$, the resulting dimensional values of zonal surface current speed from (7.28) and total circumpolar current transport (7.29) are $0.1\,\mathrm{m\,s^{-1}}$ and $100 \times 10^6\,\mathrm{m^3\,s^{-1}}$, respectively. These values are comparable to observed speeds and transports of the Antarctic Circumpolar Current. It must be emphasized, of course, that though this analytical model is a plausible, heuristically motivated solution of the large-scale dynamical balances, it is not a unique solution of a specific boundary value problem. In addition, if geostrophic motions at smaller space and time scales are allowed, the model circumpolar geostrophic jet with vertical isotherms is certain to be unstable to baroclinic disturbances so that the pure zonal parallel flow will not be observed in an uncontrolled physical system. Nonetheless, this model provides an accessible, quantitative demonstration of essential planetary geostrophic mechanisms that are likely involved in setting the basic large-scale circulation patterns of the Antarctic Circumpolar Current and the Southern Ocean.

## 7.5  Circumpolar Flow in Sverdrup Balance

The purely zonal flow of the circumpolar current, through the unblocked zonal channel, is an unrealistic feature of the previous model. A combination of wind and thermal forcing can be constructed so that the circumpolar jet has a meridional component of motion that is in Sverdrup balance with a vertical velocity induced by surface Ekman transport divergence. With the assumption of the existence of a boundary layer that closes the meridional flow, the resulting generalization provides a model for the circumpolar flow in the Southern Ocean that is substantially more representative of the observed geometry.

Suppose that the imposed surface temperature distribution is modified so that the isotherms trend southeast with constant meridional slope $dy/dx = -s < 0$:

$$T_s(x, y) = \begin{cases} T_-, & 0 \le y < \tilde{y}_-(x), \\ T_- + G_T(y - \tilde{y}_-), & \tilde{y}_-(x) \le y < \tilde{y}_+(x), \\ T_+, & y > \tilde{y}_+(x), \end{cases} \qquad (7.30)$$

where $\tilde{y}_-(x) = y_- - s(x - x_E)$, $\tilde{y}_+(x) = y_+ - s(x - x_E)$. The isotherms are assumed vertical in the current, as before, so the pressure field is given by the

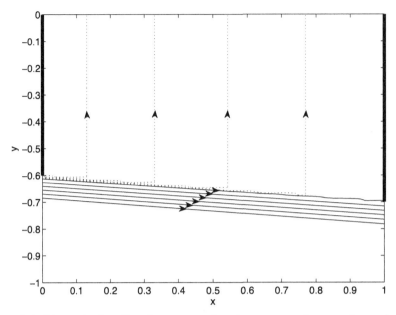

Figure 7.4. Dimensionless Sverdrup stream function contours (contour interval: 0.5, solid lines; 0.05, dotted lines) vs. longitude $x$ and latitude $y$ for the model (7.30)–(7.35) of a thermal circumpolar current in Sverdrup balance with divergent Ekman transport. The meridional stream function contours (dotted lines) represent meridional Ekman transport supported by Ekman upwelling in the circumpolar current.

same vertical integral; in this case, there is meridional as well as zonal geostrophic flow:

$$p'(y, z) = T_s(x, y)(z + H_s), \tag{7.31}$$

$$u(y, z) = -\frac{g\alpha_T G_T}{f}(z + H_s), \tag{7.32}$$

$$v(y, z) = \frac{g\alpha_T s G_T}{f}(z + H_s) \tag{7.33}$$

for $\tilde{y}_- < y < \tilde{y}_+, \quad z > -H_s$ (Figure 7.4). Thus $v/u = -s$, so the flow is along isotherms. This means that the flow again satisfies the adiabatic thermodynamic equation (4.10) because, also, $\partial T/\partial z = 0$ by assumption; indeed, (4.10) will be satisfied for any steady geostrophic flow in which $\mathbf{u} = 0$ at a fixed level and $\partial T/\partial z = 0$ because the sum of the horizontal advective terms then vanishes identically, as does the vertical advective term.

The associated vertical motion in the current can be obtained either from the Sverdrup balance or the continuity equation, giving

$$w = s\, G_T \frac{\beta}{2f^2}(z + H_s)^2, \quad \tilde{y}_-(x) \leq y < \tilde{y}_+(x), \tag{7.34}$$

so that the implied surface Ekman suction velocity is

$$W_E = s G_T \frac{\beta H_s^2}{2f^2} \equiv W_s > 0. \qquad (7.35)$$

The additional Ekman transport required to balance this surface divergence for $\tilde{y}_-(x) \leq y < \tilde{y}_+(x)$ is

$$V_{Es}(x, y) = \int_{\tilde{y}_-}^{y} W_s \, dy' = s \, G_T \frac{\beta H_s^2}{2} \left( \frac{1}{\tilde{f}_-} - \frac{1}{f} \right), \qquad (7.36)$$

where $\tilde{f}_- = f(\tilde{y}_-)$. To complete the solution for the flow in the region $y > \tilde{y}_+$ in a convenient manner, this Ekman transport can be held constant in the region $\tilde{y}_+(x) \leq y \leq y_+$ (Figure 7.4) and added at $y = y_+$ to the northward transport that is pumped downward through Ekman convergence in the subtropical gyre at latitudes $y > y_+$.

Because $\tilde{y}_-(x_E) = y_-$ and $\tilde{y}_+(x_E) = y_+$, the current flows through the gap $y_+ < y < y_-$ at $x = x_E$, as before, with $s = W_s = 0$. Now, however, it emanates from the interval $\tilde{y}_-(x_W) < y < \tilde{y}_+(x_W)$ at $x = x_W$, where $\tilde{y}_-(x_W) = y_- + s(x_E - x_W) > y_-$ and $\tilde{y}_+(x_W) = y_+ + s(x_E - x_W) > y_+$. If $s$ is sufficiently large that $\tilde{y}_-(x_W) > y_+$, this interval will lie completely north of the gap, and a barrier may be placed just downstream of the gap so that no zonal contours (except, perhaps, those in a narrow break in the barrier, to preserve the geostrophic constraint) are unblocked without obstructing the current. It is assumed that a boundary layer can exist that connects the two portions of the current by the required meridional deflection in the narrow region between the gap and the barrier. To close the flow, an amount of fluid equal to the total Ekman surface divergence in the southeastward flowing current must also be added to the current in this western boundary layer from the southward-flowing boundary current associated with the subtropical gyre in the region $y > \tilde{y}_+(x_W)$.

At each longitude $x$, the total zonal transport $\tilde{U}_c(x)$ of this current is

$$\tilde{U}_c(x) = g \alpha_T G_T \frac{H_s^2}{2\beta} \ln \frac{\tilde{f}_-}{\tilde{f}_+} \approx -g \alpha_T (T_+ - T_-) \frac{H_s^2}{2\tilde{f}_+}, \qquad (7.37)$$

where $\tilde{f}_- = f(\tilde{y}_-)$ and $\tilde{f}_+ = f(\tilde{y}_+)$. Thus the total zonal transport of the eastward current decreases gradually eastward as the Ekman suction extracts fluid from the current and exports it northward to the subtropical gyre. Although the transport $\tilde{U}_c(x)$ depends again only on thermal and geometric parameters, the eastward-decreasing portion of the transport, which arises directly in response to the Ekman suction, can be viewed as an explicitly Sverdrup-driven component of the circumpolar current, with magnitude

$$U_{\text{Sver}} = \tilde{U}_c(x_W) - \tilde{U}_c(x_E)$$

$$= g\alpha_T G_T \frac{H_s^2}{2\beta} \ln\left[\frac{f_+ \tilde{f}_-(x_W)}{f_- \tilde{f}_+(x_W)}\right]$$

$$= g\alpha_T G_T \frac{H_s^2}{2\beta} \ln\left\{\frac{y_- [y_+ + s(x_E - x_W)]}{y_+ [y_- + s(x_E - x_W)]}\right\}, \tag{7.38}$$

and is exactly equal to the integral of the Ekman suction velocity (7.35) over the total area of the current. This portion of the current does not complete a circumpolar circuit but instead returns within the basin $x_W < x < x_E$ through the subtropical gyre and the western boundary current.

Thus, when a meridional Sverdrup flow is included in this simple model, the total circumpolar transport remains independent of the Sverdrup-driven component. This example illustrates that even when the circumpolar current is in Sverdrup balance with an Ekman suction field, the Sverdrup transport balance (3.35) determines its meridional transport but not its zonal transport. Note that in this model, the imposed surface temperature distribution (7.30) and the Ekman forcing (7.35) that supports the meridional Sverdrup flow cannot be specified independently.

### 7.6 Abyssal Circulation

The models discussed in the preceding sections illustrate that in the presence of the circumpolar gap, the surface Ekman and thermodynamic forcing combine with the geometric constraints to support a baroclinic circumpolar current that flows through the gap. Associated with this current is a deep layer of warm fluid north of the gap that extends downward to the depth $z = -H_s$ of the topographic sill in the circumpolar gap. In the closed single-hemisphere basin without the circumpolar gap, a diffusively driven deep flow beneath the wind-driven subtropical gyres results from the inclusion of a turbulent heat diffusivity in the thermodynamic energy equation (Chapter 5). In the same way, turbulent heat (and salt) diffusion at the interface between the deep, warm layer north of the circumpolar gap and the abyssal, cold ($T = T_-$) layer of fluid beneath will result in a diffusively driven circulation in the abyssal layer.

A simple model of this diffusively driven abyssal circulation can be constructed by imposing a diffusively driven vertical velocity field at the top of the abyssal layer. Such an approach relies on the assumption that, as in the case of the internal boundary layer theory for the subtropical main thermocline, the diffusion is sufficiently weak that it does not alter the large-scale spatial structure of the warm-layer thickness or the layer-interface depth. Because the characteristic thickness $H_0 - H_s \approx (2/3)H_0$ of the abyssal layer is substantially larger than the wind-driven deformation $h(x, y) - H_s$ of its upper interface, the additional approximation may be made that the abyssal layer has constant thickness $H_- \approx H_0 - H_s$. If the imposed diffusively driven vertical velocity field is $W_-(x, y)$, then the associated meridional geostrophic transport $V_-$ in the abyssal layer is

$$V_- = \frac{f}{\beta} W_-. \tag{7.39}$$

Because $W_- > 0$, the abyssal transport will be poleward, just as in the case of the diffusively driven deep flow in the closed single-hemisphere basin. For the southern hemisphere basin, this transport will be toward the southern polar source of cold ($T = T_-$) fluid, and a deep western boundary current will again be required to close the circulation. The corresponding geostrophic pressure in the abyssal layer is

$$p'_-(x, y) = \frac{\rho_0 f}{H_-} \int_{x_E}^{x} V_-(x', y) dx' = \frac{\rho_0 f^2}{\beta H_-} \int_{x_E}^{x} W_-(x', y) dx', \tag{7.40}$$

from which the zonal transport $U_-$ can also be determined:

$$U_- = -\frac{1}{\rho_0 f} \frac{\partial p'_-}{\partial y}. \tag{7.41}$$

This diffusively driven circulation in the abyssal layer beneath the level of the circumpolar gap provides a simple, illustrative model of the basic dynamics of the observed abyssal meridional overturning cell associated with the formation and northward movement of Antarctic Bottom Water, its subsequent upwelling to intermediate depths at midlatitudes, and its southward return flow and renewal. There is observational evidence that deep, turbulent vertical diffusion is elevated in regions overlying relatively rough seafloor topography. Stronger diffusion and thus larger $W_-$ would lead to stronger meridional flow in such regions, with zonal flow dominating in regions of weak turbulent vertical diffusion.

## 7.7 Notes

The model of depth-integrated circumpolar flow with linear frictional closure was developed by Gill (1968). The baroclinic model of circumpolar flow is that described by Samelson (1999), which was in turn motivated by the early primitive-equation numerical experments of Gill and Bryan (1971). Theories of circumpolar flow involving the interaction of a barotropic or equivalent barotropic current with seafloor topography have received considerable attention, following the initial work of Kamenkovich (1960); the discussion here focuses on the basic large-scale thermal and geostrophic velocity structures arising from the constraints of large-scale dynamics, thermodynamics, and basin geometry rather than on the steering of the resulting circumpolar current by topographic interaction.

# 8

# Mid-Depth Meridional Overturning

## 8.1 Meridional Overturning Circulations

An essential element of the circulation of the Earth's ocean is the large-scale overturning flow that spans its full meridional reach, from the high southern to the high northern latitudes. To represent this flow, the single-hemisphere basin must be extended across the equator so that it reaches both the high southern and the high northern latitudes. Such a double-hemisphere rectangular basin, again with a circumpolar connection at the high southern latitudes, may be considered a simple model of the Atlantic sector of the world ocean (Figure 8.1). The extreme southern and northern latitude of the model basin, $y = y_S$ and $y = y_N$, will now be in the southern and northern hemisphere, respectively. The western and eastern boundaries remain at $x = x_W$ and $x = x_E$, with periodic boundary conditions for the circumpolar connection at $y_- < y < y_+, z > -H_s$.

This idealized basin retains the basic structure of the Atlantic basin, including the important circumpolar connection in the Southern Ocean. Despite neglecting many potentially significant features, including the second existing connection through the Bering Strait and the Arctic and Pacific Oceans as well as all the other complexities of seafloor topography and coastal and island geometry, it provides a useful starting point for the exploration of fundamental aspects of the large-scale overturning circulation. In such a simplified two-hemisphere basin, the circulation patterns that are naturally restricted to a single hemisphere, such as the subtropical and subpolar gyres and the circumpolar current, can be anticipated to retain the basic features and character that they possess in a single-hemisphere basin. The essential new possibility that is introduced in the two-hemisphere geometry is that of large-scale meridional flow in the presence of low surface temperatures at both latitudinal extremes of the domain, through or past a region of high surface temperatures in the intermediate midlatitude and tropical regions. This contrasts with the single-hemisphere case, in which the large-scale flow occurs in the presence of meridional profiles of large-scale surface

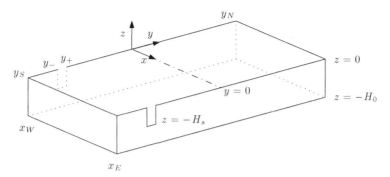

Figure 8.1. Domain geometry for the rectangular, double-hemisphere basin with circumpolar connection. The lateral boundaries are vertical, rigid walls at $x = x_E$, $x = x_W$, $y = y_S$, and $y = y_N$, except in the gap region $y_- < y < y_+$, $-H_s < z < 0$, where zonally periodic boundary conditions are imposed at $x = x_E$ and $x = x_W$, allowing circumpolar flow. The rigid, flat top and bottom are at $z = 0$ and $z = -H_0$, respectively, and the equator is at $y = 0$.

temperature that are essentially monotonic, with low temperatures at the polar extreme and high temperatures at the tropical extreme.

In this geometry, consider a conceptual extension of the analytical models of circulation in the southern hemisphere basin with circumpolar connection, in which a simple representation of observed cooling at extreme northern latitudes is included by imposing a conversion of warm fluid to cold fluid in the northern subpolar gyre. The inclusion of this surface cooling and deep-water formation at the extreme northern latitudes results in the possibility of an entirely new circulation pattern: a mid-depth meridional overturning cell that is driven by upwelling and surface warming of cold deep water at the extreme southern latitudes, as in the previous model, but that now extends across the equator to high northern latitudes, where the warm upwelled surface waters may cool, sink, and return southward to the extreme southern latitudes. This pattern represents a plausible conceptual model of the observed mid-depth cell associated with the formation and circulation of North Atlantic Deep Water. The mid-depth cell is the third basic, large-scale, zonally integrated, meridional overturning circulation pattern found in the world ocean, along with the wind-driven subtropical-centered Ekman cells and the diffusively driven abyssal cell associated with the formation of Antarctic Bottom Water. Both latter cells function in nearly the same way in the single- and double-hemisphere basin; the mid-depth cell, however, exists only in the double-hemisphere configuration.

A defining characteristic of this mid-depth meridional overturning circulation is that it allows net heat transport by the ocean from one hemisphere to the other, with surface warming of cold, upwelled water in the southern hemisphere, followed by surface cooling in the northern hemisphere. In contrast, in the single-hemisphere model discussed earlier, with a thermal circumpolar current supported by wind-driven upwelling south of the circumpolar connection and downwelling in the southern hemisphere subtropical gyre, the implied ocean heat fluxes balance locally in zonal and vertical averages so that there is no net meridional heat transport by the cell. The double-hemisphere mid-depth cell plays an important, though still poorly understood, role in the Earth's climate system, largely through its ability to support this large-scale, meridional, cross-hemisphere heat transport.

Mechanistically, the mid-depth meridional overturning cell can be viewed as a pump and valve system. There are two primary pumps that may operate: wind-driven Ekman transport northward across the circumpolar gap and interior turbulent diffusion from breaking internal waves and other sources of small-scale vertical mixing. Both mechanisms convert cold, dense water into warm, less dense water and force the warmed fluid up the surface dynamic pressure gradient across the circumpolar current or into the upper thermocline of the subtropical gyres. The reciprocal conversion of warm fluid to cold fluid in the high northern latitudes acts as a valve that controls the flux out of the warm layer, where the surface dynamic pressure is maximum.

When the valve is on, the warm–cold conversion balances the cold-warm conversion, and the fluid is cold and the surface dynamic pressure is low in the northern hemisphere subpolar gyre. In this case, a cross-equatorial overturning cell exists, with northward flow of warm surface water and southward return flow of cold deep water. Most of the meridional transport is carried by western boundary currents, both in the warm upper layer and in the cold deep layer; continuity of boundary currents and pressure gradients across the equator is assumed. When the valve is off, the model reverts essentially to the single-hemisphere configuration: there is no warm–cold conversion at high northern latitudes, the entire basin north of the gap fills with warm fluid until it reaches the sill depth, and the overturning cell is short-circuited in the southern hemisphere, with no motion north of the Ekman downwelling latitudes of the southern hemisphere subtropical gyre. Note that the valve-off southern hemisphere overturning cell still involves heat exchange in the Ekman layer and in the deep boundary current along the sill as the fluid crosses the gap; however, as in the single-hemisphere case, these exchanges balance locally in zonal and vertical averages so that there is no net meridional heat transport by the cell.

## 8.2  A Model of the Warm-Water Branch

The full meridional overturning circulation—with its enormous range of scales of active fluid motion, its interaction with seafloor and coastline topography and geometry and with the mesoscale eddy field, and its subtle dependence on small-scale turbulent flows in boundary layers and in the interior—presents a geophysical fluid dynamical problem of daunting complexity. However, some important characteristics of the large-scale flow can be exhibited and explored with a simple, one-layer, reduced-gravity model of the upper limb of the mid-depth meridional overturning cell

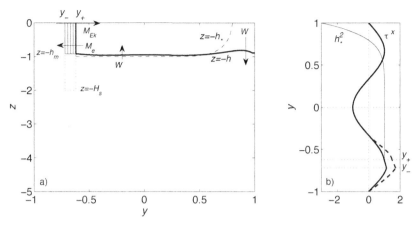

Figure 8.2. The reduced-gravity model (8.1)–(8.9), with (8.30)–(8.31) of the warm-water branch of mid-depth meridional overturning. (left) Schematic cross section showing model warm-layer depth $z = -h$ (thick solid line) and equivalent diabatic forcing depth $z = -h_*$ (dashed line) vs. dimensionless latitude $y$ and depth $z$. The model circumpolar current flows through a circumpolar gap $y_- < y < y_+$, $-H_s < z < 0$ (dotted line), at which the eastern and western boundary conditions are periodic and which extends beneath the current to the sill depth $z = -H_s$. The vertical isopycnals of the model circumpolar current associated with the warm layer (thin solid lines) fill the upper portion of the circumpolar gap to the depth $z = -h_m$, which lies above the sill. There are three types of transport into and out of the warm layer (arrows): the northward near-surface Ekman transport $M_E$ and southward interior eddy flux $M_e$ across the gap, and the vertical diabatic flux $W(x, y)$ across the base of the warm layer; the latter typically includes both upwelling at midlatitudes and downwelling at high northern latitudes. (right) Dimensionless forcing functions vs. latitude $y$. A modified stress profile with $\tau_1 = 0.5$ is also shown (dashed line). (From Samelson [2009]. © American Meteorological Society. Used with permission.)

(Figure 8.2). This upper limb is the warm-water branch of northward-flowing fluid, which is supplied by a combination of wind-driven upwelling, northward Ekman transport, and surface warming in the high-latitude southern hemisphere and diffusion-driven warming and upwelling in both hemispheres. The base of the model layer may be loosely identified with an isothermal surface near 7°C in the Atlantic, which outcrops at the latitudes of the Antarctic Circumpolar Current in the south and the northern subpolar gyre in the north, and nowhere in between. Two essential simplifications of the model are that the layer is taken to have a uniform temperature and that motions and pressure gradients in the fluid beneath the layer are neglected. Exchange of fluid between the active layer and the fluid beneath, owing to surface cooling and sinking or to diffusive warming and upwelling, is represented by a diabatic velocity $W$ at the base of the layer; this velocity also effectively includes any surface mass exchange through evaporation and precipitation.

The mass, or volume, balance of the isothermal layer with layer thickness $h(x, y, t)$ is given by the equivalent of (4.68), with the sum of the local rate of thickness change and the divergence or horizontal transport in the layer being equal to the diabatic velocity $W$ at the base of the layer:

$$\frac{\partial h}{\partial t} + \frac{\partial(hu)}{\partial x} + \frac{\partial(hv)}{\partial y} = W(x, y, t). \tag{8.1}$$

In contrast to (4.68), the Ekman pumping velocity $W_E$ does not appear in (8.1); instead, the wind stress will be included directly in the momentum balance for the layer, and the Ekman flow will be subsumed into the motion of the layer. The diabatic velocity $W$ is defined as positive upward so that $W > 0$ corresponds to a flux of fluid into the layer. For simplicity, solutions will be considered in which the layer has positive thickness ($h > 0$) everywhere in the domain north of the circumpolar current, and the thickness is also taken to vanish ($h = 0$) immediately south of the northern edge of the circumpolar current which, for simplicity, is taken to be zonal, at $y = y_+$. Thus the layer covers $x_W \leq x \leq x_E$, $y_+ \leq y \leq y_N$.

The diabatic velocity $W$ primarily represents gain or loss of fluid through the isothermal surface represented by the base of the active layer owing to small-scale turbulence, air-sea fluxes, and other small-scale processes that are not strictly large-scale phenomena but must be included because of their significant mean effect on the large-scale warm-water balance. A simple formulation, adequate to illustrate basic aspects of the role of these diabatic exchanges, is chosen here, in which the diabatic velocity $W$ is locally proportional to the difference of the square of the layer thickness $h$ and an imposed function $h_*^2(x, y, t)$, which can be chosen to represent heating at midlatitudes and cooling near the poles:

$$W(x, y, t) = -A_W[h^2(x, y, t) - h_*^2(x, y, t)]. \tag{8.2}$$

In (8.2), $A_W$ is a given constant of proportionality, and $h_*^2$ is a given function of $x$, $y$, and $t$. Note that this formulation is equivalent to the sum of a diabatic forcing, or mass source, $F_w = A_W h_*^2(x, y)$ and a nonlinear damping, or mass sink, $D_w = -A_W h^2(x, y)$. Because it is only the departure from local balance that drives the motion, however, interpretation is simpler for the difference form (8.2). In the solutions considered here, $h_*^2$ will be independent of $x$.

Equation (8.1) may be integrated over the full area of the layer to obtain the integrated, time-dependent warm-water balance. With no-normal-flow boundary conditions at the rigid walls, this gives

$$\frac{d}{dt} \int_{y_+}^{y_N} \int_{x_W}^{x_E} h \, dx \, dy = M_E + M_e + M_W. \tag{8.3}$$

In (8.3), $M_W$ is the area-integrated diabatic flux,

$$M_W = \int_{y_+}^{y_N} \int_{x_W}^{x_E} W(x, y, t) \, dx \, dy; \tag{8.4}$$

$M_E$ is the zonally integrated northward Ekman transport across the northern edge of the circumpolar current,

$$M_E = V_E^+(x_E - x_W), \quad V_E^+ = -\frac{\tau^x(y_+)}{\rho_0 f_+}, \tag{8.5}$$

where $V_E^+$ is the vertically integrated Ekman transport per unit longitude at $y = y_+$; and $M_e$ is an eddy flux of warm water across the circumpolar current. In (8.5), $f_+ = f(y_+)$, and it is assumed that the zonal wind stress at $y = y_+$ is independent of the zonal coordinate $x$.

The eddy flux $M_e$ of warm-water volume across the circumpolar current is mediated by mesoscale motions and thus, like the eddy fluxes considered in the context of the potential vorticity homogenization theory, is strictly not a large-scale circulation phenomenon. There are two primary motivations for the inclusion of this flux in the balance. First, numerous pieces of observational evidence suggest that this flux is likely a significant component of that balance. For example, it was argued in the previous chapter that in the absence of such fluxes, the circumpolar current isotherms would extend vertically downward to the sill depth, but many of the analogous isosurfaces in the Antarctic Circumpolar Current are observed, instead, to bend equatorward before the sill depth is reached. Second, as was noted in the previous chapter, the vertical-isotherm structure derived from large-scale considerations for the circumpolar current can be expected to be violently unstable to baroclinic disturbances, so it is desirable to include, at a minimum, some estimate of the mean effect that these disturbances will have on the large-scale flow.

A suitable, simple estimate of the amplitude of the eddy fluxes from such disturbances can be constructed on the assumption that their strength is locally proportional to the effective mean slope of isothermal surfaces across the current. For a circumpolar current with fixed width $L_{cc} = y_+ - y_-$, this slope can be estimated as $h(y = y_+)/L_{cc}$. The vertically integrated flux $V_e$ will then be proportional to $h^2(y = y_+)/L_{cc}$. With this motivation, the zonally and vertically integrated eddy flux $M_e$ can be taken as

$$M_e = V_e(x_E - x_W), \quad V_e = -A_e h_m^2, \tag{8.6}$$

where

$$h_m^2(t) = \frac{1}{x_E - x_W} \int_{x_W}^{x_E} h^2(x, y_+, t) \, dx \tag{8.7}$$

is the zonal-mean squared depth of the warm layer at $y = y_+$, and the factor of $1/L_{cc}$ has been absorbed into the eddy-flux coefficient $A_e$. The prescription (8.6) gives a southward eddy volume flux, out of the warm-water layer, that increases with increasing thickness of the warm layer along $y = y_+$.

To close the problem for the layer thickness $h$, it is necessary to consider the horizontal momentum equations. As in the case of the wind-driven single-hemisphere circulation, these consist of the large-scale geostrophic balance, supplemented by the Ekman transport relations. Thus a presumption of the formulation is that the warm-water branch of the mid-depth cell lies above the main, internal thermocline and is subject to the direct action of wind, at least in the sense of ventilated thermocline theory, which here is represented in its most simplistic form by a single active layer. To obtain equations that can be solved in the closed basin, it is necessary, as in the case of Equation (3.38) describing the depth-integrated wind-driven circulation in the single-hemisphere basin, to include some representation of frictional effects in the dynamical equations. As in that case, it is sufficient to add a linear frictional drag force to the depth-integrated momentum balance, which supports the presence of the lateral boundary layers that allow solution of the equations in closed basins. Here the linear friction also supports the movement of fluid across the equator in a simple equatorial boundary layer. Thus the resulting set of depth-integrated zonal and meridional momentum equations is

$$- fvh = -\gamma_1 h \frac{\partial h}{\partial x} - rhu + \frac{\tau_w^x}{\rho_0} \tag{8.8}$$

$$fuh = -\gamma_1 h \frac{\partial h}{\partial y} - rhv + \frac{\tau_w^y}{\rho_0}. \tag{8.9}$$

In (8.8)–(8.9), the Coriolis parameter is $f = \beta y$; $\gamma_1 = g\alpha_T(T_1 - T_2)$ is the reduced gravity based on the temperature difference between the warm-water layer and the fluid beneath, $r$ is a constant friction coefficient, and $(\tau_w^x, \tau_w^y) = \boldsymbol{\tau}_w$ is the imposed vector wind stress.

At the equator, a singularity arises in the planetary geostrophic equations as a result of the vanishing of the Coriolis parameter $f$. Because of this singularity, the circulation in the equatorial region can be anticipated to acquire a boundary layer character. In general, because it materially conserves the potential vorticity $f \, \partial T / \partial z$, planetary geostrophic flow is not capable of transporting flow from one hemisphere to the other, except perhaps in the special case in which the fluid is not stratified: changing the sign of $f$ while conserving potential vorticity requires changing the sign of the vertical gradient of temperature, which generally results in gravitational instability. Thus, from the point of view of large-scale theory, the equatorial boundary layer tends to act as a rigid barrier to cross-equatorial motion, with cross-equatorial exchange primarily confined to boundary sublayers at the western and eastern boundaries. As in the case of the western boundary layers for the single-hemisphere theory, it will generally be assumed for the analytical solutions that the sublayer is capable of supporting the transports required by the large-scale flow, and the details of the boundary layer structure will generally not be explored. At the eastern boundary, the primary assumption will be that the pressure, or isothermal surface depth, is continuous across the equator so that the midlatitude eastern boundary conditions will extend across the equator in a simple manner.

The equations may be made dimensionless using the dimensional scales of Section 2.1, with the depth scale $H_* = 1000$ m replacing $D$, wind stress scale $\tau_* = 0.1$ N m$^{-2}$, horizontal velocity scale $U = \tau_*/(\rho_0 f_0 H_*) = 10^{-3}$ m s$^{-1}$, $f_0 = 10^{-4}$ s$^{-1}$, and $\beta = 2 \times 10^{-11}$ m$^{-1}$ s$^{-1}$. The scale for volume transport is then $U H_* L = 5 \times 10^6$ m$^3$ s$^{-1}$. The resulting dimensionless equations have the same form as the dimensional equations, with the parameters $\gamma_1$, $h_*^2$, $A_e$, $A_W$, and $r$ scaled by $f_0 U L/H_*$, $H_*^2$, $U/H_*$, $U/(H_* L)$, and $f_0$, respectively. For $\gamma_1 = g\alpha_T \Delta T_s$, this gives a dimensionless value $\gamma_1 = 20$. The depth scale $H_*$ is a large-scale, midlatitude, warm-water penetration depth estimated from observations (Figures 1.4–1.7); a theoretically motivated alternative would be the advective thermocline depth scale $D_a$ (5.20).

## 8.3 Weak Friction and Diabatic Forcing

For steady forcing and constant parameters, a simplified, approximate form of the equations (8.1)–(8.9) may be obtained that is accurate for sufficiently small values of the friction and diabatic forcing parameters, $r$ and $A_W$, and is readily solved analytically. This simplified form retains the slow time dependence associated with the warm-water volume balance (8.3) and yields illuminating solutions for both steady and time-dependent adjustment problems.

As $r \to 0$, with $\tau_w^y = 0$, the no-normal-flow condition and (8.9) together imply that the eastern boundary depth $h_E$ of the warm layer is constant along the boundary and thus can depend only on time $t$:

$$h(x = x_E, y, t) = h_E(t). \tag{8.10}$$

Similarly, as $r \to 0$, the momentum equations (8.8)–(8.9) may be divided by $f$ and substituted into the mass conservation equation (8.1), yielding the modified, nonlinear long planetary wave equation

$$h_t - \frac{\beta\gamma_1}{f^2}hh_x + A_W h^2 = A_W h_*^2 - W_E, \tag{8.11}$$

where, as earlier,

$$W_E = \frac{1}{\rho_0}\left[\frac{\partial}{\partial x}\left(\frac{\tau_w^y}{f}\right) - \frac{\partial}{\partial y}\left(\frac{\tau_w^x}{f}\right)\right]. \tag{8.12}$$

Now, write $h^2$ as

$$h^2(x, y, t) = h_E^2(t) + D_1^2(x, y), \tag{8.13}$$

with $D_1^2(x, y)$ a function satisfying $D_1^2(x_E, y) = 0$ but otherwise remaining to be determined, and substitute this expression into (8.11) to obtain

$$\frac{h_E}{h}\frac{dh_E}{dt} - \frac{\beta\gamma_1}{2f^2}\frac{\partial D_1^2}{\partial x} + A_W h_E^2 + A_W D_1^2 = A_W h_*^2 - W_E. \tag{8.14}$$

When the flow is steady, the rate of change of $h_E$ vanishes from (8.14). In this case, the function $D_1^2(x, y)$ is determined by the steady state balance

$$\frac{\partial D_1^2}{\partial x} - \lambda D_1^2 = \lambda(h_E^2 - h_*^2) + \frac{2f^2}{\beta\gamma_1} W_E, \tag{8.15}$$

where

$$\lambda = \frac{2f^2}{\beta\gamma_1} A_W = \frac{2\beta A_W}{\gamma_1} y^2. \tag{8.16}$$

When the flow is time dependent, $dh_E/dt$ will in general be no greater, and often much smaller, than $H_*/t_{\text{adv}} = H_* U/L = 1000$ m/160 yr $= 7$ m yr$^{-1}$, while typical midlatitude values of $W_E$ are of order 30 m yr$^{-1}$. It follows that even in the time-dependent case, it is consistent to neglect $dh_E/dt$ in (8.14) so that the function $D_1^2(x, y)$ is determined to first order by the steady state balance (8.15). This is equivalent to the assertion that the time scale for adjustment of $h_E$ is long compared to the basin-crossing time at the long planetary wave speed $c_R = \beta\gamma_1 h/f^2$. If, in addition, $A_W$ is small enough that $\lambda(x_E - x_W) \ll 1$, the terms in (8.15) that are proportional to $\lambda$ may also be neglected, yielding

$$h(x, y, t) = \left[h_E^2(t) + D_1^2(x, y)\right]^{1/2} \tag{8.17}$$

$$D_1^2(x, y) = \frac{2f^2}{\beta\gamma_1} \int_{x_E}^{x} W_E(x', y)\, dx'. \tag{8.18}$$

Thus the instantaneous spatial structure of $h(x, y, t)$ is rationally approximated in this limit by the familiar reduced-gravity expression of the steady state, adiabatic, wind-driven Sverdrup theory. However, $h$ depends on the still unknown instantaneous value $h_E(t)$ of the eastern boundary thickness.

The condition that determines $h_E(t)$ can be obtained from the time-dependent integrated warm-water volume balance (8.3). The rate of change of warm-water volume is

$$\frac{d}{dt}\int_{y_+}^{y_N}\int_{x_W}^{x_E} h\, dx\, dy = h_E \frac{dh_E}{dt}\int_{y_+}^{y_N}\int_{x_W}^{x_E} \frac{1}{h}\, dx\, dy. \tag{8.19}$$

The latter integral depends on $h_E$ and is not easily evaluated analytically. An adequate approximation for $h$ near the given characteristic thickness $H = 1000$ m of the warm-water layer is

$$\int_{y_+}^{y_N}\int_{x_W}^{x_E} \frac{1}{h}\, dx\, dy \approx \frac{1}{h_E}(x_E - x_W)(y_N - y_+); \tag{8.20}$$

that is, the effective average of the wind-driven deformations $D_1^2(x, y)$ over the warm-layer extent may, in this equation, be taken to be small relative to the eastern boundary thickness $h_E$. If, also, $\partial(\tau_w^x/f)/\partial y \approx 0$ at $y = y_+$, so that $h_m^2 = h_E^2$, then $M_e$ depends only on $h_E$:

$$M_e = -(x_E - x_W)A_e h_E^2. \tag{8.21}$$

With $h$ given in terms of $h_E$ and $D_1^2$, the northward transport of warm water per unit longitude at latitude $y$ is

$$V(y) = -\int_y^{y_N} W(y')\,dy' = A_W(y_N - y)\left[h_E^2 + \bar{D}_1^2(y) - \bar{h}_*^2(y)\right], \tag{8.22}$$

where

$$
\begin{aligned}
\bar{D}_1^2(y) &= \frac{x_E - x_W}{y_N - y}\int_y^{y_N}\frac{f^2}{\beta\gamma_1}\frac{\partial}{\partial y}\left(\frac{\tau_w^x}{f}\right)dy \\
&= \frac{x_E - x_W}{y_N - y}\left\{\frac{1}{\gamma_1}\left[(y\tau_w^x)\big|_y^{y_N} - 2\int_y^{y_N}\tau_w^x\,dy\right]\right\}
\end{aligned}
\tag{8.23}
$$

and

$$\bar{h}_*^2(y) = \frac{1}{y_N - y}\int_y^{y_N} h_*^2(y')\,dy' \tag{8.24}$$

are known functions of $y$. Here it has been assumed that $\tau_w^y = 0$ and that $\tau_w^x$ is independent of $x$. Then $M_W$ is given by

$$
\begin{aligned}
M_W &= -(x_E - x_W)V(y_+) \\
&= -A_W(x_E - x_W)(y_N - y_+)[h_E^2 + \bar{D}_1^2(y_+) - \bar{h}_*^2(y_+)],
\end{aligned}
\tag{8.25}
$$

which depends only on known functions and on the unknown eastern boundary thickness $h_E$.

With the approximations (8.20), (8.21), and (8.25) to the corresponding terms in (8.3), the resulting equation for the evolution of $h_E$ is

$$\frac{dh_E}{dt} + \frac{A_e + A_W(y_N - y_+)}{y_N - y_+}h_E^2 = \frac{V_E^+}{y_N - y_+} - A_W[\bar{D}_1^2(y_+) - \bar{h}_*^2(y_+)]. \tag{8.26}$$

Thus the time-dependent dynamics of this model reduce in this limit to the single, first-order, ordinary differential equation (8.26). Although Equation (8.26) is nonlinear, it has constant coefficients, and its analytical steady and time-dependent solutions can be readily derived.

## 8.4 Steady Solutions

The steady solutions $h_{Es}$ of (8.26) may be obtained by setting $dh_E/dt = 0$ and $h_E = h_{Es}$ and solving for $h_{Es}^2$. This yields

$$h_{Es}^2 = \frac{V_E^+ - A_W(y_N - y_+)\left[\bar{D}_1^2(y_+) - \bar{h}_*^2(y_+)\right]}{A_e + A_W(y_N - y_+)}. \tag{8.27}$$

The corresponding eddy and integrated diabatic volume fluxes are

$$M_e = -(x_E - x_W)A_e h_{Es}^2 \tag{8.28}$$

$$M_W = -(x_E - x_W)V(y_+) = -(x_E - x_W)\left[V_E^+ - A_e h_{Es}^2\right]. \tag{8.29}$$

The condition $dh_E/dt = 0$ means that the net flux of mass, or volume, into the warm-water layer vanishes; that is, $M_E + M_e + M_W = 0$. This steady solution is obtained when the eastern boundary depth is such that the area-integrated diabatic flux $M_W$ exactly balances the sum of the Ekman and eddy transports, $M_E$ and $M_e$, at the southern boundary of the layer, at $y = y_+$, along the northern edge of the circumpolar current. In general, $M_E$ will be positive, representing a northward Ekman transport; $M_e$ will be negative, representing a southward eddy flux; and $M_W$, which represents an integrated, net flux over upwelling and downwelling regions, may take either sign.

For the specific solutions discussed here, the imposed wind stress $\boldsymbol{\tau}_w$ is taken to be purely zonal (Figure 8.2), with

$$\frac{\tau_w^x}{\rho_0} = \begin{cases} \tau_0 \cos\frac{3\pi y}{2}, & y_3 < y < y_N = 1, \\ \tau_0 \cos\frac{3\pi y}{2} + \frac{1}{2}\tau_1\left(1 - \cos\frac{\pi(y - y_3)}{y_+ - y_3}\right), & y_+ \le y < y_3 < 0, \end{cases} \tag{8.30}$$

and $\tau_w^y = 0$, and the squared equivalent thickness function $h_*^2(y)$ for the diabatic forcing north of the circumpolar channel is given by

$$h_*^2(y) = \begin{cases} h_0^2 - \delta h_N^2\, y^6, & 0 < y < y_N = 1, \\ h_0^2, & y_+ < y < 0, \\ 0, & y_S < y < y_+. \end{cases} \tag{8.31}$$

The wind-stress coeffficients $\tau_0$ and $\tau_1$ and the midlatitude equivalent diabatic forcing depth $h_0$ may be chosen to represent observed annual and zonal mean zonal wind stress profiles and the approximate maximum depth of the subtropical main thermocline, respectively. Note that $h_*^2$ may be negative for sufficiently large $\delta h_N^2$ and $y$ (Figure 8.2); by (8.2), this guarantees local cooling and diabatic flux out of the warm layer ($W < 0$) at high northern latitudes because the squared physical layer depth $h^2$ can never be negative.

For general values of the parameters, steady solutions of the model equations (8.1)–(8.9), with (8.30)–(8.31), may also be obtained numerically. In this case, it is

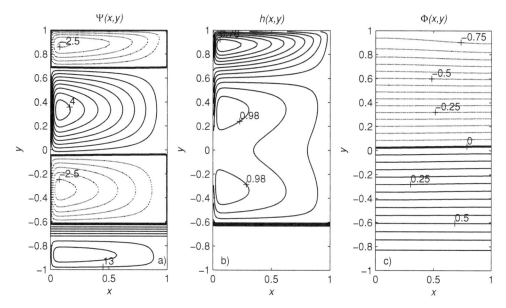

Figure 8.3. Contours vs. $x$ and $y$ for a steady dimensionless numerical solution of (8.1)–(8.9), with (8.30)–(8.31) and $(A_W, A_e) = (2, 1)$, $\tau_0 = -1$, $\tau_1 = 0$, $h_0 = 1$, $\delta h_N^2 = 4$, $\gamma = 20$, $r = 0.02$, and $\beta = 1$: (a) $\Psi$ (contour interval (CI) = 0.5, with CI = 2 in $y_- = -0.72 < y < y_+ = -0.62$ and CI = 1 for $y < y_-$), (b) $h$ (CI = 0.02; $h = 0$ for $y < y_+$), and (c) $\Phi$ (CI = 0.05). (From Samelson [2009]. © American Meteorological Society. Used with permission.)

convenient to decompose the horizontal transport into components from a stream function $\Psi$ and a potential function $\Phi$ so that

$$hu = -\frac{\partial \Psi}{\partial y} - \frac{\partial \Phi}{\partial x}, \quad hv = \frac{\partial \Psi}{\partial x} - \frac{\partial \Phi}{\partial y}. \tag{8.32}$$

For weak friction and diabatic forcing, these numerical solutions have the form anticipated by the asymptotic analysis. The stream function and thickness fields show the familiar Sverdrup structure, with thickness increasing westward in the anticyclonic subtropical gyres and decreasing westward in the cyclonic northern subpolar gyre (Figures 8.3a and 8.3b). Superimposed on this wind-driven structure is the diabatic meridional flow, represented by down-gradient flow from the transport potential (Figure 8.3c). The meridional flow is supported by northward Ekman transport across the circumpolar current, diabatic upwelling at midlatitudes, and downwelling in the northern subpolar gyre (Figure 8.4). The balance between Ekman transport and midlatitude upwelling as sources for the warm-water meridional overturning circulation depends on the values of the parameters, especially $\tau_0$, $\tau_1$, $A_W$, and $A_e$ (Figure 8.5).

The most realistic representations of the ocean's mid-depth circulation are likely provided by solutions in which the Ekman and upwelling sources contribute roughly equal parts to the overturning transport. Appropriate values of the eddy flux and

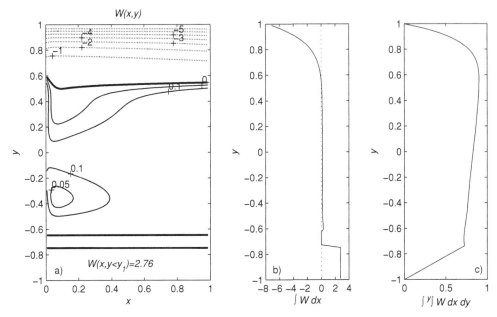

Figure 8.4. (a) Contours of $W$ vs. $x$ and $y$ (CI $= 1$ for $W < 0$ and CI $= 0.05$ for $W > 0$) and zonally integrated (b) vertical velocity $\int_{x_W}^{x_E} W(x, y) \, dx$ and (c) northward transport $-\int_{x_W}^{x_E} \Phi_y(x, y) \, dx$ vs. $y$ for the steady numerical solution in Figure 8.3. For this solution, $W(x, y) = 0$ for $y_- < y < y_+$, and $W(x, y) = 2.76$ for $y_S = 0 < y < y_-$. (From Samelson [2009]. © American Meteorological Society. Used with permission.)

diabatic forcing constants $A_e$ and $A_W$ are otherwise uncertain. The analytical solution shows that in general, the eastern boundary depth $h_E$ is not strongly sensitive to variations in $A_e$ and $A_W$ when $A_e + A_W \geq 1$. As both $A_e$ and $A_W$ approach zero, however, $h_E$ approaches infinity. Thus, for sufficiently weak eddy activity and diabatic forcing, the warm water will extend downward to the sill depth of the circumpolar channel. Note that for the preceding scales, $\lambda \leq 0.1 \, A_W$, so the analytical solution should be useful for, roughly, $A_W \leq 3$. Comparisons with numerical solutions confirm this and show further that the analytical solution can provide useful guidance even for larger values of $A_W$ (Figure 8.5c).

From the analytical solution, other relations between parameters can be easily obtained. For example, a special set of solutions can be found for which there is complete compensation of the Ekman transport by the eddy flux so that $M_W = V(y_+) = 0$. For variable $\tau_0$ and $\delta h_N^2$, for any $A_W > 0$, and with other parameter values fixed as earlier, the eddy coefficient $A_e$ that gives this compensation varies over the range $0 < A_e < 8$ and is approximately proportional to $\tau_0$ for fixed $\delta h_N^2$. Thus complete shutdown of the wind-driven component of the overturning can be achieved for parameter values that are not extreme. Note that in the analytical limit of small $\lambda$, these values of $A_e$ are independent of $A_W$.

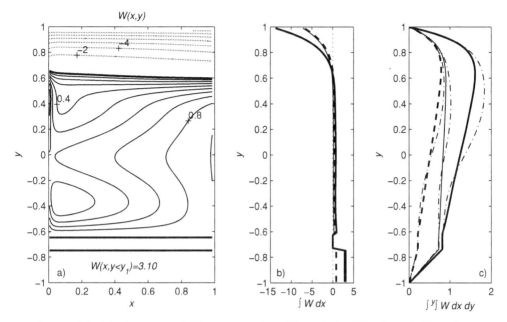

Figure 8.5. (a) Contours of $W$ vs. $x$ and $y$ (CI = 1 for $W < 0$ and CI = 0.1 for $W > 0$) for a steady dimensionless numerical solution of (8.1)–(8.9), with $(A_W, A_e) = (4, 1)$ and other parameters as in Figure 8.3; for this solution, $W(x, y) = 3.10$ for $y_S < y < y_-$. Zonally integrated (b) vertical velocity $\int_{x_W}^{x_E} W(x, y)\, dx$ and (c) northward transport $-\int_{x_W}^{x_E} \Phi_y(x, y)\, dx$ vs. $y$ for steady numerical solutions with $(A_w, A_e) = \{(4, 1), (2, 1), (2, 2)\}$ (thick solid line, thin solid line, and thick dashed line, respectively) and other parameters as in Figure 8.3. In (c), the corresponding approximate analytical results for the zonally integrated northward transport are also shown (dash-dotted lines). (From Samelson [2009]. © American Meteorological Society. Used with permission.)

As might be anticipated, the eddy flux $M_e$ depends strongly on $A_e$ and only weakly on $A_W$. However, as $A_W \to 0$, the dependence of $M_e$ on $A_e$ weakens for larger values of $A_e$. In this regime, $A_e$ has exceeded the value that shuts down the wind-driven overturning, and the net flow meridional flow $V(y_+)$ across the circumpolar current is southward. The small values of $A_W$ hinder upwelling of additional warm fluid north of the current, leading to the saturation of $M_e$ as $A_e$ increases. In contrast, the net meridional flow $V(y_+)$ across the channel is, perhaps surprisingly, nearly independent of $A_W$, except for $A_W \leq 1/2$, for which the shutdown of diabatic exchange north of the channel results in approximate compensation for any $A_e$.

Variations in the parameter $\tau_1$ affect only the southern hemisphere winds near the circumpolar gap. The dependence of the net northward transport across the current, $V(y_+)$, on $\tau_1$ is strong for most values of $A_e$ and $A_W$, weakening slightly for large $A_e$. For small values of $A_W$, the dependence of $V(y_+)$ on $\tau_1$ also weakens, as the eddy fluxes remove the warm-layer fluid more efficiently than the diabatic forcing, resulting in weak, approximately compensated transport across the current. In contrast, for fixed $A_e = 1$, the eastern boundary depth $h_E$ generally depends only weakly on $\tau_1$ for fixed $A_W$. The dependence of the eddy flux $M_e$ on $\tau_1$ is relatively strong for large $A_e$ and small $A_W$ and relatively weak for small $A_e$ and large $A_W$. The meridional transport $V(y)$ north of the circumpolar gap and in the northern hemisphere also depends on $\tau_1$. At the equator ($y = 0$), this dependence is diminished roughly by half from the dependence of $V(y_+)$ on $\tau_1$. In the central latitudes of the northern hemisphere subtropical gyre, this dependence is reduced by roughly half again.

Although highly idealized, the model illustrates several essential aspects of the dynamics of the mid-depth meridional overturning. The amplitude and distribution of the warm-water transport is determined by a three-way balance of northward Ekman transport across the circumpolar current, southward eddy fluxes across the circumpolar current, and diabatic exchange north of the circumpolar current, with the latter typically including midlatitude upwelling and high northern latitude down-welling. The model also illustrates the critical role of the eastern boundary layer depth $h_E$ in setting the thermocline structure north of the current. Because this depth is communicated instantaneously, in this approximation, along the eastern boundary, it effectively controls the mean difference between the squared layer depth $h^2$ and the squared equivalent diabatic forcing depth $h_*^2$ throughout the warm-water layer and thus also the net diabatic flux through the base of the warm layer. At the southern edge of the warm layer, this same depth controls the strength of the eddy fluxes across the circumpolar current as well as the zonal transport of the current itself. Thus the eastern boundary depth is an essential mediator in the competition between eddy fluxes and interior diabatic fluxes that determines the amplitude of the overturning cell.

## 8.5 Time-Dependent Solutions

The time-dependent equation (8.26) for the eastern boundary thickness $h_E(t)$ is non-linear. However, for steady forcing and constant parameters, it may be readily solved. The time-dependent general solution is

$$h_E(t) = h_{Es} + \Delta h_E \left\{ \frac{2h_{Es} \exp(-2\mu_A h_{Es}t)}{2h_{Es} + \Delta h_E[1 - \exp(-2\mu_A h_{Es}t)]} \right\}, \qquad (8.33)$$

where $h_{Es}$ is the steady solution (8.27); the constant $\Delta h_E$ is the difference of the initial value of $h_E(t)$ from $h_{Es}$,

$$\Delta h_E = h_E(0) - h_{Es}; \qquad (8.34)$$

and the parameter $\mu_A$ is given by

$$\mu_A = \frac{A_e + A_W(y_N - y_+)}{y_N - y_+}. \qquad (8.35)$$

Thus the time-dependent solution approaches the steady solution exponentially, with decay time scale

$$T_{MOC} = \frac{1}{2\mu_A h_{Es}}. \tag{8.36}$$

The approximate analytical solution (8.33)–(8.35) has several immediate implications. First, under conditions of steady forcing and parameter values, the approach to the unique steady state solution is monotonic; despite the nonlinearity, there are no oscillations and no multiple equilibria. Second, the nonlinearity does cause a dependence of the approach time scale $T_{MOC}$ (8.36) on the steady state solution value of the eastern boundary thickness such that the approach to solutions with relatively larger eastern boundary thicknesses will be relatively faster. Third, and perhaps most important, the approach time scale $T_{MOC}$ is directly proportional to the eddy flux and diabatic parameters $A_e$ and $A_W$.

Note also that whereas (8.33)–(8.35) describe the intrinsic time scale of thermocline adjustment associated with the mid-depth meridional overturning, the adjustment of the northward flux of warm water that comprises the warm-water branch of the mid-depth overturning may occur more rapidly. This is because of the strong control exerted on the meridional flux of warm water by changes in external forcing or parameters: the effect on $W$ of an instantaneous change in $h_*$ may be comparable in magnitude to that from the long-term adjustment of $h$, and the effect on $M_E$ of an instantaneous change in $\tau_1$ may generally be larger than the change in $W$ or $M_e$ from the resulting adjustment of $h$. Thus the warm-water branch of meridional overturning may respond much more rapidly than the thermocline structure to changes in surface forcing.

For general parameter values and time-dependent forcing, the initial boundary value problem for the full model equations (8.1)–(8.9) may be solved numerically by standard methods. Steady solutions may be obtained numerically either by time stepping the time-dependent equations or by solving the steady problem directly. Comparison with the numerical solutions shows that the analytical approximations are useful even for moderate values of the friction and diabatic flux coefficients. These comparisons can be made directly for solutions that are obtained by initializing the model with a known steady state, abruptly changing the forcing or parameters to a different value, and integrating to equilibrium with the new forcing or parameter values held fixed (Figure 8.6).

For the time-dependent problem, the numerical solutions give additional insight into the short-term adjustment processes that are neglected in the analytical approximation. With abruptly changed forcing or parameters, the time-dependent adjustment from the initial to the final state proceeds in several stages. On very short time scales $t \approx T_{\text{local}} \ll t_{\text{adv}}$, with $T_{\text{local}}$ corresponding to dimensional times of several months, there is a local response in (8.1) to direct forcing or parameter changes, which often extend southward along the western boundary to the equator and eastward along

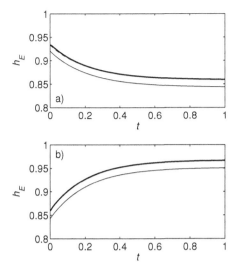

Figure 8.6. Eastern boundary thickness $h_E$ vs. time $t$ from dimensionless numerical solutions (thick solid lines) and from the analytical solution (8.33)–(8.35) (thin solid lines). Results for a time evolution described by the functional form (8.33)–(8.35), but with initial and final values taken from the numerical solutions, are indistinguishable from the numerical solutions. For the numerical solutions, the meridional mean of $h_E(y, t)$ is shown; the corresponding standard deviation of $h_E(y, t)$ from the meridional mean at each time was 0.002 or less. In each case, the solutions were initialized at steady states similar to those illustrated in Figures 8.3–8.5. The model diabatic or wind-forcing parameters were then changed abruptly at time $t = 0$ and were held fixed until the solutions approached the new equilibrium. (From Samelson [2011]. © American Meteorological Society. Used with permission.)

the equator of the leading edge of the disturbance. These changes are essentially instantaneous relative to the advective time scale $t_{adv} \approx 160$ years. Note that the time-dependent penetration along the western boundary and equator may include boundary-advective effects and is not simply Kelvin wave propagation; in the numerical solutions with finite friction parameter $r$, the damped analog of Kelvin wave propagation is effected instantaneously by the solution of a differential equation around the boundary that results from the no-normal-flow condition.

On the longer, but still short, time scales ($t \approx T_{pwave} \approx L/c_R \ll t_{adv}$) of planetary wave basin-crossing times, corresponding to dimensional times of several years to a decade, the familiar zonal slope of the layer thickness associated with the Sverdrup balance (8.18) is established. The pathway of the signal is eastward along the equator to the eastern boundary, then poleward along the eastern boundary, and then westward via planetary waves. Finally, on the full ($t \approx T_{MOC} \approx t_{adv}$) adjustment time scale, the mean layer thickness evolves toward the equilibrium state, while the horizontal structure of the layer thickness variations remains roughly constant. This adjustment is essentially an exponential decay and is well described by the analytic solution (8.33)–(8.35), with a dimensional time scale typically corresponding to multiple decades or longer. As is shown explicitly by the analytic solution, it is the eddy and diabatic flux parameters, $A_e$ and $A_W$, along with the steady solution itself, that determine this adjustment time scale.

The latter result is of particular note and is consistent with the related dependence of the steady solution (8.27) on the same parameters. This dependence of the long-term adjustment time scale on parameters of the representation of small-scale motions on the large-scale, planetary geostrophic flow is at once illuminating and disappointing: it illustrates and emphasizes the essential dependence of the large-scale flow on the mean effect of fluctuating, small-scale processes and thereby reveals another aspect of the unfortunate incompleteness of the purely large-scale asymptotic theory, along with the mathematical degeneracy of the planetary geostrophic equations. For the model that yielded (8.27) and (8.36), the essential mathematical degeneracy has been addressed by introducing the linear friction, proportional to the parameter $r$; happily, the results (8.27) and (8.36) are independent of that parameter. However, as illustrated by (8.27) and (8.36), the essential physical balances controlling the adjustment time scale of the large-scale circulation and thermal structure in this model prove to be those associated with the representations of small-scale eddy flux and diabatic processes—not at the lateral boundaries but in the ocean interior—rather than the large-scale dynamical balances themselves.

## 8.6 Notes

Early theoretical models of meridional overturning circulation focused on control of diabatic upwelling velocities in the subtropical thermocline by turbulent diffusion (Stommel and Webster, 1962). Control of mid-depth meridional overturning by southern hemisphere Ekman transport across the circumpolar current was first proposed by Toggweiler and Samuels (1995) on the basis of numerical experiments. Scaling theories for the thermocline structure associated with the three-way balance of Ekman transport, eddy fluxes across the circumpolar current, and diabatic vertical fluxes were subsequently developed by Gnanadesikan (1999) and Johnson et al. (2007), following Welander (1971) and more general analyses by Tziperman (1986) and de Szoeke (1995) of the control exerted on thermocline structure and eastern-boundary layer thicknesses by integral diabatic flux balances. The analytical models discussed in this chapter follow Samelson (2004, 2009, 2011).

# 9

# Thermohaline Effects

## 9.1 Seawater as a Two-Component Fluid

All the theoretical models of large-scale circulation discussed in the preceding chapters have as their starting point the simplified planetary geostrophic equations (4.7)–(4.10), in which the independent temperature, salinity, and density variables have been replaced by a single thermodynamic variable. In contrast, the existing theory for large-scale circulation based on the more general thermohaline planetary geostrophic equations (2.102)–(2.107), with independent temperature and salinity variables and a nonlinear equation of state for density, is severely limited.

There are two primary reasons for the lack of theoretical progress on the thermohaline planetary geostrophic equations. The first reason is that the inclusion of an independent salinity variable leads to a general mathematical problem of considerably greater intrinsic difficulty, involving coupled partial differential equations for two independent scalar fields. For the general nonlinear equation of state (1.19), the hydrostatic balance (2.26) in the vertical momentum equation must be computed as

$$\frac{\partial p}{\partial z} = -g\rho = -g\tilde{\mathcal{R}}(p, \theta, S). \tag{9.1}$$

To obtain the potential temperature and salinity fields, $\theta$ and $S$, that are necessary to evaluate $\tilde{\mathcal{R}}$, it is necessary to solve the evolution equations for both $\theta$ and $S$. Through (9.1) and the geostrophic balance (2.102), the velocity field thus depends on both of the two advected fields: temperature and salinity. This means, equivalently, that a general reduction of this system to a single partial differential equation for a single scalar field, analogous to the single equation (4.57) for the simplified planetary geostrophic equations, is not available. An evolution equation may be computed for density,

$$\frac{D\rho}{Dt} = \frac{\partial \tilde{\mathcal{R}}}{\partial p} \frac{Dp}{Dt} + \frac{\partial \tilde{\mathcal{R}}}{\partial \theta} \frac{D\theta}{Dt} + \frac{\partial \tilde{\mathcal{R}}}{\partial S} \frac{DS}{Dt}, \tag{9.2}$$

but in general, this leads to no essential simplification because the equation is not closed in terms of the density $\rho$.

The second reason is that ironically, despite this additional intrinsic complexity, the adiabatic, nondiffusive forms (2.107) and (2.105) of the planetary geostrophic thermodynamic energy and salinity equations invite a degeneracy of solutions that is not easily transcended. For steady solutions, the material conservation statements (2.107) and (2.105) for potential temperature and salinity reduce to

$$\mathbf{u} \cdot \nabla\theta = \mathbf{u} \cdot \nabla S = 0. \tag{9.3}$$

This means both that salinity and temperature must be constant along each fluid trajectory and that the flow must be confined within surfaces of constant $\theta$ and constant $S$. If the flow on these surfaces contains recirculating regions, the situation is therefore similar to that considered in the theory of potential vorticity homogenization: the inclusion of weak lateral diffusion might be anticipated to lead to the homogenization of $S$ on $\theta$ surfaces, and of $\theta$ on $S$ surfaces, and thus to a functional dependence of one variable on the other. Note that the restricted two-dimensional flow on $\theta$ and $S$ surfaces implied by (9.3) need not be incompressible on the respective surfaces; instead, the corresponding two-dimensional velocity fields $\mathbf{u}_\theta$ and $\mathbf{u}_S$ satisfy

$$\nabla_h \cdot \left(\mathbf{u}_\theta \frac{\partial Z}{\partial \theta}\right) = \nabla_h \cdot \left(\mathbf{u}_S \frac{\partial Z}{\partial S}\right) = 0, \tag{9.4}$$

where $\partial Z / \partial \theta$ and $\partial Z / \partial S$ are the differential thicknesses of the surfaces of constant $\theta$ and constant $S$, respectively. Thus the flow kinematics are analogous to those for the planetary geostrophic homogenization theory, in which a transport, rather than velocity, stream function could be found for the two-dimensional flow.

The analogy to homogenization theory is not exact, however, in part because the diffusion need not be confined within the respective isothermal and isohaline surfaces. Diffusion across the surfaces will induce advective flow normal to the surfaces, while the homogenization theory assumes strictly two-dimensional flow. Nonetheless, the analogy suggests a general physical tendency of the large-scale flow to develop temperature-salinity relations in regions of recirculation. At the same time, the strict requirements (9.3) strongly constrain solutions in the steady, adiabatic theory as both salinity and temperature must be constant along each fluid trajectory. Consequently, theoretical models based on the thermohaline equations that have solutions with substantially independent temperature and salinity fields generally must include representations of diabatic and nonconservative effects, further complicating the mathematics of the model. In most cases, the resulting equations can only be solved using numerical methods.

## 9.2 Thermohaline Surface Boundary Conditions

For the linear equation of state (4.2) with constant $\alpha_T$ and $\beta_S$, it may be possible to write the evolution equation for density without separate reference to temperature and salinity, as discussed in Section 4.1. For example, consider a formulation including vertical turbulent diffusion in the thermodynamic energy and salinity equations,

$$\frac{DT}{Dt} = \kappa_v \frac{\partial^2 T}{\partial z^2}, \quad \frac{DS}{Dt} = \kappa_v \frac{\partial^2 S}{\partial z^2}, \tag{9.5}$$

with equal turbulent diffusivities $\kappa_v$. These evolution equations for $T$ and $S$ are identical in form, and the density anomaly $\rho' = -\rho_0(\alpha_T T - \beta_S S)$ will therefore satisfy the same evolution equation:

$$\begin{aligned}
\frac{D\rho'}{Dt} &= \rho_0 \left( -\alpha_T \frac{DT}{Dt} + \beta_S \frac{DS}{Dt} \right) \\
&= \rho_0 \kappa_v \left( -\alpha_T \frac{\partial^2 T}{\partial z^2} + \beta_S \frac{\partial^2 S}{\partial z^2} \right) = \kappa_v \frac{\partial^2 \rho'}{\partial z^2}.
\end{aligned} \tag{9.6}$$

In this case, the interior equations are closed in terms of the density anomaly $\rho'$. However, unless the thermal and haline boundary conditions can also be expressed in terms of $\rho'$, it will still be necessary to retain independent temperature and salinity variables and evolution equations in the thermohaline planetary geostrophic equations (4.7)–(4.10).

The surface boundary conditions on temperature and salinity reflect different physical processes, and in many cases, they will take differing forms. The general thermal boundary condition is a heat flux condition, including sensible, latent, and radiative fluxes, the first two of which depend strongly on the local air-sea temperature difference. The natural boundary condition on salinity is a freshwater flux condition, with volume flux equal to the difference of precipitation and evaporation, which is essentially independent of the local surface salinity. For large-scale theory, highly simplified forms of these boundary conditions may be sufficient, such as the direct specification of surface temperature (5.3) or the relaxation toward an imposed surface air temperature (5.4). Consistent with the Boussinesq incompressibility and rigid-lid conditions, the freshwater flux condition is often replaced with an effective salinity flux, representing the effect of the freshwater flux on the local salinity.

Suppose that with interior flow described by a single density equation such as (9.6), the surface salinity and surface temperature are both specified directly so that the surface boundary conditions take the form (5.3) and

$$S(x, y, 0, t) = S_a(x, y, t), \tag{9.7}$$

respectively, where $S_a$ is a given surface salinity function. These boundary conditions may be expressed as a single boundary condition on density anomaly:

$$\rho'(x, y, 0, t) = \rho_0[-\alpha_T T_a(x, y, t) + \beta_S S_a(x, y, t)]. \qquad (9.8)$$

Similarly, if an effective surface salinity flux is imposed along with the surface heat flux (5.4), with identical relaxation time scale $\gamma_a^{-1}$, so that the surface heat and effective salinity fluxes are given by (5.4) and

$$Q_s(x, y, t) = -\gamma_a[S(x, y, 0, t) - S_a(x, y, t)], \qquad (9.9)$$

respectively, then the boundary condition may again be expressed in terms of an equivalent density (or mass) flux,

$$Q_\rho(x, y) = \gamma_a[\rho'(x, y, 0, t) - \rho_a(x, y, t)], \qquad (9.10)$$

where $\rho_a = \rho_0(-\alpha_T T_a + \beta_S S_a)$. In either of these cases, with no-flux conditions on $T$ and $S$ at the remaining rigid boundaries, the full problem is closed in terms of density alone. The use of a single density variable, as in the theories discussed in the preceding chapters, is then consistent, as the partitioning of the density fluctuations into thermal and haline contributions is not required for any aspect of the solution.

If, however, the boundary conditions for temperature and salinity are such that either the surface temperature or salinity fields must be individually available, then a single density variable is not sufficient. Such boundary conditions often arise because of the different physical processes determining the surface fluxes. For example, the dependence of heat flux on the local air-sea temperature difference suggests that the temperature boundary condition should include a contribution of the form (5.4), while the lack of dependence of the freshwater flux on local air-sea differences suggests that the salinity condition may be better represented as a given function of space and time:

$$Q_s(x, y, t) = F_S(x, y, t). \qquad (9.11)$$

In this case, though the imposed freshwater or effective salinity flux function $F_S(x, y, t)$ has no dependence on the local surface salinity, it is necessary to know the surface temperature field to compute the surface thermal boundary condition. Consequently, the replacement of temperature and salinity by a single density variable is not possible, even if a single density equation, such as (9.6), can be derived for the interior flow. Instead, the temperature and salinity fields retain their dynamical independence, and the corresponding evolution equations—(9.5) or their equivalents—must both be solved. The idealized model described in the next section provides an illustration of the additional complexity that can arise, even under highly constrained flow conditions, from this independence, including the intriguing possibility of self-sustaining large-scale thermohaline oscillations.

## 9.3 Multiple Equilibria for a Thermohaline Exchange Flow

Considerable complexity can arise in the thermohaline dynamics when external forcing or internal mixing induces independent variations in temperature and salinity, even for the simplified case of a linearized equation of state. This complexity can be illustrated with a highly idealized model of zonally integrated meridional overturning, in which the circulation is represented by the interaction of two independent water types in two well-mixed reservoirs. In this case, independent temperature and salinity variables must be retained because of differing dynamical constants in the corresponding thermodynamic energy and salinity equations.

Let the temperature and salinity in reservoir $j$, $j = \{1, 2\}$ be denoted by $T_j$ and $S_j$, respectively. Suppose that the dynamics and forcing are such that the reservoir temperatures and salinities relax toward the externally imposed values $T_{jr}$ and $S_{jr}$ at different rates $\gamma_T$ and $\gamma_S$, respectively. A difference in these rates can be anticipated from the different physical processes controlling the fluxes of heat and freshwater through the boundaries of the reservoir, for example, from the surface flux processes described earlier, or, in a laboratory setting, from reservoir walls constructed of membranes that pass heat more freely than salt. Suppose also that the reservoirs are connected by a capillary tube below and a free-surface overflow above, with equal and opposite flow $V$ in the capillary and $-V$ in the overflow. The volume flux $V$ from reservoir 1 to reservoir 2 is taken to be proportional to the difference $\rho_1 - \rho_2$ in the respective densities, where density is given by the linear equation of state (4.2). For antisymmetric forcing, $T_{1r} = -T_{2r} = T_r > 0$ and $S_{1r} = -S_{2r} = S_r > 0$, antisymmetric solutions may be sought, which will have $T = T_1 = -T_2$ and $S = S_1 = -S_2$. For these solutions, the thermodynamic energy and salinity equations reduce to

$$\frac{dT}{dt} = \gamma_T(T_r - T) - 2|V|T \tag{9.12}$$

$$\frac{dS}{dt} = \gamma_S(S_r - S) - 2|V|S, \tag{9.13}$$

where

$$V = k_q(\rho_1 - \rho_2) = 2k_q\rho_0(-\alpha_T T + \beta_S S) \tag{9.14}$$

and $k_q$ is a constant of proportionality. Note that in this simple model, in which the reservoirs are instantaneously mixed, the exchange of heat and salt induced by the circulation is independent of the direction of the circulation: for either $V > 0$ or $V < 0$, fluid of temperature $T$ is removed from reservoir 1 and fluid of temperature $-T$ is added in its place.

Even with the linear equation of state, two independent variables representing temperature $T$ and salinity $S$ must be retained to describe the system (9.12)–(9.14).

For, if $\gamma_T \neq \gamma_S$, Equations (9.12) and (9.13) are independent even when $V = 0$: a single closed equation cannot be formed for any linear combination of $T$ and $S$, including the density deviation $\rho - \rho_0 = \rho_0(-\alpha_T T + \beta_S S)$. The different time-dependent relaxation dynamics for $T$ and $S$ in this model, which, in a more complete model, might be related to differing boundary flux properties for heat and freshwater, are sufficient to prevent the replacement of $T$ and $S$ by a single density variable, even though the equation of state is linear.

This simple model has an additional property of interest: for fixed $T_r$ and $S_r$, and fixed values of the parameters, it may have more than one steady solution. Steady solutions of (9.12)–(9.14) must satisfy

$$T/T_r = 1/(1 + 2|V|/\gamma_T), \quad S/S_r = 1/(1 + 2|V|/\gamma_S), \tag{9.15}$$

$$V/V_0 = -\frac{1}{1 + 2|V|/\gamma_T} + R_\rho \frac{1}{1 + 2|V|/\gamma_S}, \tag{9.16}$$

where

$$V_0 = 2k_q \rho_0 \alpha_T T_r, \quad R_\rho = \frac{\beta_S S_r}{\alpha_T T_r}. \tag{9.17}$$

For given values of the parameters, the solution of (9.16) may be obtained graphically as the intersection points in the $(V, Y)$ plane of functions $Y = F_L(V)$ and $Y = F_R(V)$ that represent the left-hand and right-hand sides of the equation, respectively. The left-hand function $F_L = V/V_0$ represents a simple ray passing through the origin, with positive slope $1/V_0$, which is arbitrarily small when $V_0$ is arbitrarily large. The right-hand function $F_R$ is equal to $R_\rho - 1$ at $V = 0$ and, if $R_\rho > 1$ and $\gamma_d$ is sufficiently small, has zero crossings at $V = \pm(R_\rho - 1)/[2(\gamma_T R_\rho - \gamma_S)]$. Thus, for $V_0$ and $R_\rho$ sufficiently large and $\gamma_S$ sufficiently small, the curves must intersect at one point where $Y > 0$ and $V > 0$ and at a second point where $Y < 0$ and $V < 0$; a third intersection between these two may also occur.

Thus, for certain sets of fixed boundary ($T_r$ and $S_r$) and parameter values, this model has multiple equilibrium solutions rather than a single, unique solution. It is the combination of the differing relaxation time scales for temperature and salinity, and the advective nonlinearity that gives rise to the possibility of these multiple equilibria. If the advective term is linearized, by setting $V = $ constant in (9.12) and (9.13), the resulting equations for $T$ and $S$ are linear, independent equations with exponential solutions and unique equilibria. Similarly, if $\gamma_T = \gamma_S$, then the function $F_R(V)$ representing the right-hand side of (9.16) tends monotonically toward zero with increasing $|V|$, and only a single solution of (9.16) will exist.

Because $T > 0$ and $S > 0$, the flux relation (9.14) requires that $V < 0$ when temperature differences control density differences and $V > 0$ when salinity differences control density differences. Thus the solution with $V < 0$ has density dominated by temperature, capillary flow from the cold to the warm reservoir, and surface return flow from the warm to the cold reservoir. The solution with $V > 0$ has density dominated by salinity, capillary flow from the warm to the cold reservoir, and surface return flow from the cold to the warm reservoir. In this highly idealized model, then, the two possible equilibrium solutions represent meridional overturning circulations in opposite directions: one controlled by thermal effects and the other controlled by by haline effects. If a nonlinear dependence of flux on density difference is introduced in place of the linear flux relation (9.14), parameter regimes can be found for which these model equations support sustained, time-dependent thermohaline oscillations.

## 9.4 Thermohaline Potential Vorticity

Many of the theoretical results described in the preceding chapters relied on the availability of a conserved planetary geostrophic potential vorticity. In the thermohaline case, a general form of a planetary geostrophic potential vorticity is

$$Q_\lambda = f \frac{\partial \lambda}{\partial z}, \tag{9.18}$$

where $\lambda$ is a suitable scalar, typically a thermodynamic variable. The simplified form (9.18) of the potential vorticity $Q_\lambda$ arises because, for the dimensional planetary geostrophic scalar field $\lambda$,

$$\left| \frac{\nabla_2 \lambda}{\partial \lambda / \partial z} \right| \approx \frac{D}{L} \ll 1, \tag{9.19}$$

where characteristic values of the depth and length scales, $D$ and $L$, are given in (2.1). The anisotropic scaling of $\nabla \lambda$ eliminates contributions that would otherwise arise from the horizontal components of the planetary geostrophic approximation to the absolute vorticity, which have magnitudes comparable to the magnitude of the vertical component $f\mathbf{k}$. In the special case of the linear equation of state (4.2) with salinity $S \equiv 0$, the choice $\lambda = T$ may be made, and the planetary geostrophic material conservation law (4.49) then holds for the corresponding potential vorticity $Q_T$.

In general, however, no planetary geostrophic potential vorticity (9.18) is materially conserved by thermohaline planetary geostrophic flow. The material derivative of $Q_\lambda$ is

$$\frac{DQ_\lambda}{Dt} = \left( \beta v - f \frac{\partial w}{\partial z} \right) \frac{\partial \lambda}{\partial z} + f \frac{\partial}{\partial z} \left( \frac{D\lambda}{Dt} \right) + \frac{g}{\rho_0} \left( \frac{\partial \rho}{\partial x} \frac{\partial \lambda}{\partial y} - \frac{\partial \rho}{\partial y} \frac{\partial \lambda}{\partial x} \right). \tag{9.20}$$

In view of the planetary geostrophic vorticity relation (3.8), the first term on the right-hand side of (9.20) vanishes identically. In general, the second term will vanish if (and only if) a material conservation law holds for $\lambda$, that is, if

$$\frac{D\lambda}{Dt} = 0, \tag{9.21}$$

whereas the third will vanish if (and only if) $\lambda$ is a function only of the density $\rho$ and depth $z$, that is, if

$$\lambda = \Lambda(\rho, z). \tag{9.22}$$

If (9.22) holds, then

$$\frac{D\Lambda}{Dt} = \frac{\partial\Lambda}{\partial\rho}\frac{D\rho}{Dt} + w\frac{\partial\Lambda}{\partial z}. \tag{9.23}$$

Then, by (9.2), with materially conserved $\theta$ and $S$,

$$\frac{D\Lambda}{Dt} = \frac{\partial\Lambda}{\partial\rho}\frac{\partial\tilde{\mathcal{R}}}{\partial p}\frac{Dp}{Dt} + w\frac{\partial\Lambda}{\partial z} \approx w\left(\frac{\partial\Lambda}{\partial z} - g\rho_0\frac{\partial\Lambda}{\partial\rho}\frac{\partial\tilde{\mathcal{R}}}{\partial p}\right), \tag{9.24}$$

where $p \approx p_0 - g\rho_0 z$ has been used because it is the full pressure that appears in the equation of state. In general, $w \neq 0$ for solutions of interest, so a materially conserved potential vorticity can only be found if a function $\Lambda(\rho, z)$ exists such that

$$\left(\frac{\partial\Lambda}{\partial\rho}\right)^{-1}\frac{\partial\Lambda}{\partial z} = g\rho_0\frac{\partial\tilde{\mathcal{R}}}{\partial p}. \tag{9.25}$$

The left-hand side of (9.25) must be independent of $\theta$ and $S$, which in turn requires that

$$\frac{\partial}{\partial\theta}\frac{\partial\tilde{\mathcal{R}}}{\partial p} = \frac{\partial}{\partial S}\frac{\partial\tilde{\mathcal{R}}}{\partial p} = 0. \tag{9.26}$$

Because, for the full equation of state, the compressibility $\partial\tilde{\mathcal{R}}/\partial p$ of seawater depends both on temperature and, to a lesser degree, on salinity, it is in general not possible to find a materially conserved potential vorticity for the full thermohaline planetary geostrophic equations.

It follows from these considerations that the conservation law (4.49) for planetary geostrophic potential vorticity must be seen as an approximation even within the planetary geostrophic framework, based on the additional assumption that density may be taken to be linearly proportional to temperature (or potential temperature). In the slightly more general case of a linear equation of state (4.2) with both temperature and salinity materially conserved, the potential vorticity (9.18) with $\lambda = \rho$ is materially conserved, following (9.6), with $\kappa_v = 0$.

Note that the origin of the third term in (9.20) is, effectively, an approximation,

$$\nabla\lambda \cdot (\nabla\rho \times \nabla p) \approx \nabla\lambda \cdot \left(\nabla\rho \times \frac{\partial p}{\partial z}\mathbf{k}\right), \tag{9.27}$$

to the vector triple product of the gradients $\nabla\lambda$, $\nabla\rho$, and $\nabla p$ that arises in the general potential vorticity equation for a compressible fluid. However, direct estimation of the full triple product, with the anisotropic scaling (9.19), does not yield the approximation (9.27) because neglected terms that are proportional to $\partial\lambda/\partial z$ or $\partial\rho/\partial z$

appear formally to be of the same order as the terms that are retained. Instead, the neglect of terms in the triple product (9.27) can be traced to the Boussinesq approximation (2.41) in the horizontal momentum equations, which removes terms of the form $\nabla \rho \times (\nabla_h p, 0)$ from the vorticity equation. In the special cases noted earlier, in which $\lambda = T$ or $\lambda = \rho$ and $Q_\lambda$ is materially conserved, the full triple product vanishes identically and the approximation remains consistent.

## 9.5 Pressure Coordinates

The planetary geostrophic circulation theory that has been developed in the preceding chapters is formulated in terms of mass conservation and momentum equations that have been simplified through the Boussinesq approximation (Section 2.4). The process of carrying out the Boussinesq approximation has the virtue of making explicit some basic aspects of the physics and thermodynamics of seawater in the context of large-scale ocean circulation. An alternative path toward simplified forms of these equations that does not rely on the Boussinesq approximation is a transformation to pressure coordinates, in which the vertical coordinate is transformed from distance $z$ to pressure $p$. This approach has the alternative appeal of making explicit a broad simplification of the equations of motion that arises from the hydrostatic approximation. Note that the Boussinesq approximation does not depend on the hydrostatic approximation and can be made instead directly on the nonhydrostatic equations, yielding a nonhydrostatic Boussinesq system that is often used, for example, to study the convective instabilities that arise when a fluid is heated from below or cooled from above. Alternatively, the pressure-coordinate transformation shows that when the hydrostatic approximation is made first, the Boussinesq approximations in the mass conservation and momentum equations are unnecessary.

Suppose that the hydrostatic balance (2.26) holds. Let $P(\lambda, \phi, z, t)$ be the pressure function in the distance vertical coordinate $z$, and consider an arbitrary function $\tilde{F}(\lambda, \phi, p, t)$ in the pressure vertical coordinate $p$, so that

$$F(\lambda, \phi, z, t) = \tilde{F}[\lambda, \phi, P(\lambda, \phi, z, t), t] \tag{9.28}$$

gives the value of the function $\tilde{F}$ in the distance vertical coordinate $z$. Then

$$\frac{\partial F}{\partial z} = \frac{\partial \tilde{F}}{\partial p} \frac{\partial P}{\partial z}, \quad \frac{\partial F}{\partial t} = \frac{\partial \tilde{F}}{\partial t} + \frac{\partial \tilde{F}}{\partial p} \frac{\partial P}{\partial t}, \tag{9.29}$$

and similar expressions to the latter hold for $\partial F / \partial \lambda$ and $\partial F / \partial \phi$, with $t$ replaced everywhere by $\lambda$ or $\phi$, respectively. Also, define

$$\omega = \frac{DP}{Dt}, \tag{9.30}$$

where $\omega$ is the material rate of change of the hydrostatic pressure. It follows, then, that

$$\frac{DF}{Dt} = \frac{D\tilde{F}}{Dt} = \frac{\partial \tilde{F}}{\partial t} + \tilde{\mathbf{u}}_h \cdot \nabla_h \tilde{F} + \omega \frac{\partial \tilde{F}}{\partial p}, \tag{9.31}$$

where $\tilde{\mathbf{u}}_h(\lambda, \phi, p, t)$ is the horizontal velocity in the pressure vertical coordinate, defined so that

$$\mathbf{u}_h(\lambda, \phi, z, t) = \tilde{\mathbf{u}}_h[\lambda, \phi, P(\lambda, \phi, z, t), t], \tag{9.32}$$

with $\mathbf{u}_h = (u^\lambda, u^\phi)$; here the two-dimensional horizontal gradient operator $\nabla_h$ is to operate with pressure $p$ held fixed because $\tilde{F}$ depends explicitly on $p$. Now, replace $\rho$ in the mass conservation equation (1.15) using the hydrostatic relation (2.26), and use (9.29) and (2.16) to obtain

$$
\begin{aligned}
0 &= \frac{D}{Dt}\left(-\frac{1}{g}\frac{\partial P}{\partial z}\right) - \frac{1}{g}\frac{\partial P}{\partial z}\nabla \cdot \mathbf{u} \\
&= -\frac{1}{g}\left(\frac{\partial}{\partial z}\frac{DP}{Dt} - \frac{\partial \mathbf{u}}{\partial z}\cdot \nabla P + \frac{\partial P}{\partial z}\nabla \cdot \mathbf{u}\right) \\
&= -\frac{1}{g}\frac{\partial P}{\partial z}\left[\frac{\partial \omega}{\partial p} + \frac{1}{R_e \cos \phi}\left(\frac{\partial u^\lambda}{\partial \lambda} - \frac{\partial \tilde{u}^\lambda}{\partial p}\frac{\partial P}{\partial \lambda}\right) + \frac{1}{R_e}\left(\frac{\partial u^\phi}{\partial \phi} - \frac{\partial \tilde{u}^\phi}{\partial p}\frac{\partial P}{\partial \phi}\right)\right] \\
&= \rho\left(\frac{1}{R_e \cos \phi}\frac{\partial \tilde{u}^\lambda}{\partial \lambda} + \frac{1}{R_e}\frac{\partial \tilde{u}^\phi}{\partial \phi} + \frac{\partial \omega}{\partial p}\right). \tag{9.33}
\end{aligned}
$$

Because the density $\rho$ by definition cannot vanish within the domain occupied by the fluid, it is thus a consequence of the hydrostatic approximation that the flow in pressure coordinates $(\lambda, \phi, p)$ is incompressible:

$$\nabla_p \cdot (\tilde{\mathbf{u}}_h, \omega) = \nabla_h \cdot \tilde{\mathbf{u}}_h + \frac{\partial \omega}{\partial p} = 0, \tag{9.34}$$

where $\nabla_p$ is the gradient operator analogous to (2.16) in pressure coordinates. Note that the approximation (2.16) has been made in the divergence, which has an error of order $D/R_e \approx 10^{-3}$.

With the hydrostatic approximation (2.26), the evolution equation for the vertical velocity $w$ is replaced by a diagnostic relation between pressure and density. Through the pressure-coordinate expression (9.34) of mass conservation, this approximation yields a diagnostic equation for the analogous pressure-coordinate vertical velocity $\omega$: the evolution equations for the horizontal velocity $\tilde{\mathbf{u}}_h$ are the horizontal momentum equations, as before, but now $\omega$ can be determined from $\tilde{\mathbf{u}}_h$ at each $t$ simply by integrating (9.34) vertically, given a suitable boundary condition. The pressure coordinates allow an additional simplification, through the introduction of the Montgomery function $\mathcal{M}(\lambda, \phi, p, t) = gZ + p/\rho_0$, where $Z(\lambda, \phi, p, t)$ is the inverse of $P$. Then, because

$$\left[\frac{1}{\rho}\nabla_h P(\lambda, \phi, z, t)\right]\Bigg|_{z=Z} = \nabla_h \mathcal{M}, \qquad (9.35)$$

the density no longer appears explicitly in the inviscid horizontal momentum equations:

$$\frac{D\tilde{\mathbf{u}}_h}{Dt} + 2\Omega \sin \phi\, \tilde{\mathbf{k}} \times \tilde{\mathbf{u}}_h = -\nabla_h \mathcal{M}. \qquad (9.36)$$

In terms of $M$, the hydrostatic balance (2.26) is

$$\frac{\partial \mathcal{M}}{\partial p} = \frac{1}{\rho_0} - \frac{1}{\rho}. \qquad (9.37)$$

The pressure-coordinate form of the hydrostatic primitive equations is (9.36), (9.37), and (9.34), along with the salinity equation (2.98) and thermodynamic energy equation and equation of state (2.99) or (2.100), where now, the material derivative is given by (9.31). The mathematical structure of these equations is the same as that of the hydrostatic, Boussinesq primitive equations (2.95)–(2.100), though the corresponding physical variables have slightly different physical meanings. However, no approximations to the mass conservation equation or the horizontal inertia have been made in (9.36) or (9.34), except for the neglect of the Coriolis term in $w$ in the zonal component of (9.36) and the approximation (2.16) to the spherical-coordinate divergence. Because the bottom boundary is at an unknown value of the vertical coordinate $p$ in pressure coordinates, rather than a known value of $z$, the specification of the bottom boundary condition takes the mathematically more difficult form of a free-boundary problem. Note that in (2.99) or (2.100), the Boussinesq approximation in the thermodynamic energy equation has been made. It is the Boussinesq approximation to the mass and momentum equations that has been avoided by the transformation to pressure coordinates, with the fundamental consequent simplification being the essentially exact incompressibility condition (9.34).

The pressure-coordinate expression of the hydrostatic primitive equations is widely used in numerical models of the atmosphere, for which the Boussinesq approximation of the mass conservation equation is much less accurate because of the much greater compressibility of air relative to that of water. It has the advantage of the near-complete removal of acoustic waves owing to the elimination from the mass conservation equation of the time derivative $\partial \rho/\partial t$ of density, a crucial element of acoustic wave dynamics. The transformation to pressure coordinates thus appears, mysteriously, to remove a second evolution equation. The density time derivative, however, effectively reappears in a modified boundary condition. In the ocean, the resulting boundary-trapped acoustic mode is only a minor modification of an existing external gravity mode. Nonetheless, the modified boundary conditions present some additional complications for ocean modeling, which has instead relied heavily on the Boussinesq mass and momentum equations.

## 9.6 Thermohaline Planetary Geostrophic Equations in Pressure Coordinates

The apparent inconsistency of the thermohaline planetary geostrophic vorticity dynamics outlined in Section 9.4 may be explicitly avoided if an analogous set of equations is formulated in pressure coordinates. In this case, the geostrophic approximation to the horizontal momentum equations is

$$f\tilde{\mathbf{k}} \times \tilde{\mathbf{u}}_h = -\nabla_h \mathcal{M},$$ (9.38)

where $\nabla_h$ is the horizontal gradient operator with pressure $p$ held fixed and, importantly, no approximation to the inertia is made. Here $\mathcal{M} = gZ(x, y, p) + p/\rho_0$ is the Montgomery function, which satisfies the hydrostatic equation

$$\frac{\partial \mathcal{M}}{\partial p} = \frac{1}{\rho_0} - \frac{1}{\rho},$$ (9.39)

with the function $z = Z(x, y, p)$ giving the standard vertical distance coordinate. The hydrostatic approximation leads directly to the solenoidal mass-conserving form (9.34) of the continuity equation,

$$\nabla_h \cdot \tilde{\mathbf{u}}_h + \frac{\partial \omega}{\partial p} = 0,$$ (9.40)

without the portion of the Boussinesq approximation that is required to obtain the incompressibility condition in vertical distance coordinates. The large-scale salinity and thermodynamic energy equations and the equation of state are, as before, (2.105) and (2.107) [or, alternatively, (2.106)], where the material derivative is now (9.31), and these, with (9.38)–(9.40), form the thermohaline planetary geostrophic equations in pressure coordinates. Note that the density dependence on pressure in the equation of state transforms into a direct dependence on the vertical coordinate $p$, which is now an independent rather than dependent variable.

In this formulation, the thermal wind equations are

$$f\tilde{\mathbf{k}} \times \frac{\partial \tilde{\mathbf{u}}_H}{\partial p} = -\frac{1}{\rho^2} \nabla_h \rho,$$ (9.41)

and the analog of the Sverdrup vorticity relation (3.8) is

$$\beta \tilde{v} = f \frac{\partial \omega}{\partial p}.$$ (9.42)

Similarly, the general potential vorticity $Q_\lambda$ has the form

$$Q_\lambda = f \frac{\partial \lambda}{\partial p},$$ (9.43)

and its material derivative is

$$\frac{DQ_\lambda}{Dt} = \left(\beta\tilde{v} - f\frac{\partial\omega}{\partial p}\right)\frac{\partial\lambda}{\partial p} + f\frac{\partial}{\partial p}\left(\frac{D\lambda}{Dt}\right) + \frac{1}{\rho^2}\left(\frac{\partial\lambda}{\partial x}\frac{\partial\rho}{\partial y} - \frac{\partial\lambda}{\partial y}\frac{\partial\rho}{\partial x}\right), \qquad (9.44)$$

where the partial derivatives in the last term are taken with $p$ held fixed. In this case, the last term in (9.44) is proportional to the full vector triple product $\nabla\lambda \cdot (\nabla\rho \times \nabla p)$ because $\nabla p = (0, 0, 1)$ in pressure coordinates. Thus the corresponding condition on the function $\lambda = \Lambda$ is that $\Lambda$ depend only on $\rho$ and $p$, exactly as for a general compressible fluid.

An argument exactly analogous to that involving (9.21)–(9.26) in the preceding section then shows that no thermohaline planetary geostrophic potential vorticity $Q_\lambda$ can be found that will be materially conserved in the general case, in which the compressibility $\partial\tilde{\mathcal{R}}/\partial p$ depends on temperature and salinity. As before, the case of a linear equation of state (4.2) is an exception for which the choice $\lambda = T$ (if no salinity variations are allowed) or $\lambda = \rho$ can give a materially conserved potential vorticity. An additional special case, in which a materially conserved potential vorticity may be defined, occurs when the solution has a fixed temperature-salinity relation, that is, if the fields $\theta$ and $S$ are related everywhere by a function $\mathcal{S}(\theta)$ such that

$$S = \mathcal{S}(\theta). \qquad (9.45)$$

In this case, the equation of state reduces to

$$\rho = \hat{\mathcal{R}}(p, \theta) = \tilde{\mathcal{R}}[p, \theta, \mathcal{S}(\theta)], \qquad (9.46)$$

and the function $\hat{\mathcal{R}}$ can be inverted for $\theta$, giving

$$\theta = \Theta(\rho, p). \qquad (9.47)$$

When $D\theta/Dt = 0$, the choice $\Lambda \equiv \theta = \Theta(\rho, p)$ then gives the materially conserved potential vorticity $Q_\theta$, where

$$Q_\theta = f\frac{\partial\theta}{\partial p}. \qquad (9.48)$$

This result can be seen from an alternative perspective, as follows. In general, the proportional change of density with infinitesimal increments of pressure, potential temperature, and salinity is given by the differential of the nonlinear equation of state from (2.107),

$$d\rho = \frac{\partial\tilde{\mathcal{R}}}{\partial p}dp + \frac{\partial\tilde{\mathcal{R}}}{\partial\theta}d\theta + \frac{\partial\tilde{\mathcal{R}}}{\partial S}dS. \qquad (9.49)$$

When a temperature-salinity relation exists, this differential reduces to

$$d\rho = \frac{\partial\tilde{\mathcal{R}}}{\partial p}dp + \left(\frac{\partial\tilde{\mathcal{R}}}{\partial\theta} + \frac{\partial\tilde{\mathcal{R}}}{\partial S}\mathcal{S}'\right)d\theta, \qquad (9.50)$$

where $S' = dS/d\theta$. This can be rearranged to obtain

$$d\theta = \left( \frac{\partial \tilde{\mathcal{R}}}{\partial \theta} + \frac{\partial \tilde{\mathcal{R}}}{\partial S} S' \right)^{-1} \left( d\rho - \frac{\partial \tilde{\mathcal{R}}}{\partial p} dp \right). \tag{9.51}$$

Because $\theta$ is a thermodynamic variable, the differential $d\theta$ is exact. Hence the first term on the right-hand side of (9.51) is an integrating factor for the second term, from which it follows from (9.51) that $\theta = \Theta(\rho, p)$, where $\Theta$ can, in principle, be obtained directly by integration.

## 9.7 A Simple Nonlinear Equation of State

The empirical function, typically a high-order polynomial, that is currently recognized as the most accurate approximation to the equation of state $\rho = \tilde{\mathcal{R}}(p, \theta, S)$ will generally be unwieldy for theoretical analysis. Simplified forms may be sought that retain the dominant nonlinearity but are more amenable to analysis. One such form is

$$\rho = \frac{\rho_0}{\mathcal{V}(p, \theta, S)}, \tag{9.52}$$

where

$$\mathcal{V}(p, \theta, S) = 1 - \frac{1}{\rho_0 c_0^2} \left( p - \frac{1}{2} \gamma_c p^2 \right) + \alpha_T (1 + \gamma p)(\theta - \theta_r)$$

$$+ \frac{1}{2} \alpha_2 (\theta - \theta_r)^2 - \beta_S (S - S_r). \tag{9.53}$$

Here $\rho_0, \theta_r$, and $S_r$ are reference density, temperature, and salinity values, respectively; $\alpha_T$ and $\alpha_2$ are the first and second thermal expansion coefficients, respectively; $\beta_S$ is the haline contraction coefficient; $\gamma$ is the thermobaric coefficient; $1/(\rho_0 c_0^2)$ is the compressibility; and $\gamma_c$ is the compressibility pressure coefficient. For the values $\rho_0 = 1027.7$ kg m$^{-3}$, $\theta_r = 5°C$, $S_r = 35$, $\alpha_T = 1.067 \times 10^{-4}$ °C$^{-1}$, $\alpha_2 = 1.041 \times 10^{-5}$ °C$^{-2}$, $\beta_S = 0.754 \times 10^{-3}$, $\gamma = 1.86 \times 10^{-8}$ Pa$^{-1}$, $c_0 = 1466$ m s$^{-1}$, $1/(\rho_0 c_0^2) = 4.53 \times 10^{-10}$ Pa$^{-1}$, and $\gamma_c = 2.98 \times 10^{-9}$ Pa$^{-1}$, this approximate equation of state has an accuracy of 0.1 kg m$^{-3}$ or better for most oceanographically relevant values of $p, \theta$, and $S$.

From (9.52), the evolution equation for density may be written as

$$\frac{D\rho}{Dt} = -\frac{\rho^2}{\rho_0} \frac{D\mathcal{V}}{Dt}, \tag{9.54}$$

where

$$\frac{D\mathcal{V}}{Dt} = \frac{\partial \mathcal{V}}{\partial p} \frac{Dp}{Dt} + \frac{\partial \mathcal{V}}{\partial \theta} \frac{D\theta}{Dt} + \frac{\partial \mathcal{V}}{\partial S} \frac{DS}{Dt} \tag{9.55}$$

and

$$\frac{\partial \mathcal{V}}{\partial p} = -\frac{1}{\rho_0 c_0^2}(1 - \gamma_c p) + \gamma \alpha_T (\theta - \theta_r), \tag{9.56}$$

$$\frac{\partial \mathcal{V}}{\partial \theta} = \alpha_T (1 + \gamma p) + \alpha_2 (\theta - \theta_r), \tag{9.57}$$

$$\frac{\partial \mathcal{V}}{\partial S} = -\beta_S. \tag{9.58}$$

Note that the partial derivative with respect to $p$ in (9.56) is taken with the thermodynamic variables $\theta$ and $S$ held fixed and must be distinguished from the partial derivative with respect to $p$ in the pressure coordinate system, which is taken with $x$ and $y$ held fixed.

Examination of (9.56)–(9.58) shows that the right-hand side of (9.54) cannot be written in terms of the density $\rho$ alone. Furthermore, even when the temperature $\theta$ and salinity $S$ are both materially conserved, the density $\rho$ is not, because of the effect $\partial \mathcal{V}/\partial p$ of pressure on density. The latter effect depends on $\theta$, so even in this case, $\theta$ must be known independently of $\rho$. To solve this system of equations, it is thus necessary to integrate the evolution equations for both temperature $\theta$ and salinity $S$ and to substitute the resulting values at each point into the equation of state (9.52), to obtain the density to be used in the hydrostatic relation. An equivalent approach is, at each time $t$, to substitute the instantaneous fields $\theta$ and $S$ into the evolution equation (9.54) for density $\rho$, and then to integrate (9.54) simultaneously with the evolution equations for $\theta$ and $S$.

For this equation of state, the arguments in the preceding sections show that it is the pressure-dependent compressibility, represented by the nonzero thermobaric coefficient $\gamma$, that prevents the construction of a materially conserved planetary geostrophic potential vorticity for the simplified equation of state (9.52). In the pressure-coordinate formulation, for example, the equivalent of (9.24) is

$$\frac{D\Lambda}{Dt} = -\frac{\rho^2}{\rho_0}\omega \left\{ \frac{\partial \Lambda}{\partial p} + \left[ -\frac{1}{\rho_0 c_0^2}(1 - \gamma_c p) + \gamma \alpha_T (\theta - \theta_r) \right] \frac{\partial \Lambda}{\partial \rho} \right\}. \tag{9.59}$$

With nonzero thermobaric effect—that is, with $\gamma \neq 0$ in (9.52)—no nontrivial thermodynamic function $\Lambda$ can be found that is dependent only on $\rho$ and $p$ and satisfies $D\Lambda/Dt = 0$, and consequently, no materially conserved planetary geostrophic potential vorticity can be found. For this equation of state—and in general—much of the theory discussed in the preceding chapters must therefore be understood as an approximation to the full thermohaline planetary geostrophic dynamics. In general, the magnitude of error admitted by the additional approximation necessary to allow the construction of a materially conserved potential vorticity will be proportional to the magnitude of the thermobaric coefficient $\gamma$ because the thermobaric effect dominates relative to other potentially complicating factors such as the dependence of the compressibility on salinity.

In contrast, if there were no thermobaric effect, that is, if

$$\gamma = 0, \tag{9.60}$$

then a pressure-corrected density $\sigma$ dependent only on $\rho$ and $p$ could be defined by

$$\sigma(\rho, p) = \rho + \rho_0 c_0^2 \left( p - \frac{1}{2}\gamma_c p^2 \right). \tag{9.61}$$

In this case, with $\gamma = 0$, the right-hand side of (9.59) would vanish identically for the choice $\Lambda(\rho, p) = \sigma(\rho, p)$ and the associated potential vorticity $Q_\sigma$, where

$$Q_\sigma = f \frac{\partial \sigma}{\partial p}, \tag{9.62}$$

would be materially conserved.

Similarly, suppose that a linear temperature-salinity relation exists so that

$$S = \mathcal{S}(\theta) = S_r + R\frac{\alpha_T}{\beta_S}(\theta - \theta_r), \tag{9.63}$$

where $R$ is a constant. Then, from (9.52), it follows that

$$\theta = \theta_r + \Theta_1(\rho, p), \tag{9.64}$$

where

$$\Theta_1(\rho, p) = -\frac{\alpha_T}{\alpha_2}(1 - R + \gamma p)$$
$$+ \left\{ \frac{\alpha_T^2}{\alpha_2^2}(1 - R + \gamma p)^2 + \frac{2}{\alpha_2}\left[ 1 - \frac{\rho_0}{\rho} - \frac{1}{\rho_0 c_0^2}\left( p - \frac{1}{2}\gamma_c p^2 \right) \right] \right\}^{1/2}, \tag{9.65}$$

so the potential temperature $\theta$ is explicitly a function of $\rho$ and $p$. Then, for the dynamic density field

$$\rho = \hat{\rho}(x, y, p, t), \tag{9.66}$$

the corresponding materially conserved potential vorticity is

$$Q_\theta(\hat{\rho}, p) = f \frac{\partial \Theta_1[\hat{\rho}(x, y, p), p]}{\partial p}$$
$$= f \left( -\gamma \frac{\alpha_T}{\alpha_2} + \frac{Q_1}{Q_2} \right), \tag{9.67}$$

where

$$Q_1 = \frac{\alpha_T^2}{\alpha_2^2}\gamma(1 - R + \gamma p)p - \frac{2}{\alpha_2}\left[\frac{\rho_0}{\hat{\rho}^2}\frac{\partial\hat{\rho}}{\partial p} - \frac{1}{\rho_0 c_0^2}(1 - \gamma_c p)\right] \qquad (9.68)$$

$$Q_2 = \Theta_1(\hat{\rho}, p) + \frac{\alpha_T}{\alpha_2}(1 - R + \gamma p). \qquad (9.69)$$

Although $Q_\theta$ may appear complicated in form, it is clear that $DQ_\theta/Dt = 0$ when $D\theta/Dt = 0$ because $Q_\theta$ depends only on $\rho$ and $p$. This is an explicit example of the general result (9.48) on the existence of a materially conserved potential vorticity when temperature and salinity are functionally related. If the temperature-salinity relation (9.63) exists and, in addition, the coefficient $\alpha_2 = 0$ in (9.52), then the corresponding potential temperature and materially conserved potential vorticity functions are

$$\Theta_1(\rho, p) = Q_3(p)\left[\frac{\rho_0}{\rho} - 1 + \frac{1}{\rho_0 c_0^2}\left(p - \frac{1}{2}\gamma_c p^2\right)\right] \qquad (9.70)$$

$$Q_\theta(\hat{\rho}, p) = f Q_3(p)\left[-\frac{\rho_0}{\hat{\rho}^2}\frac{\partial\hat{\rho}}{\partial p} + \frac{1}{\rho_0 c_0^2}(1 - \gamma_c p) - \alpha_T\gamma\Theta_1(\hat{\rho}, p)\right], \qquad (9.71)$$

where

$$Q_3(p) = \frac{1}{\alpha_T(1 - R + \gamma p)}. \qquad (9.72)$$

The approximate correspondence of the product $f(\partial\hat{\rho}/\partial p)$ in (9.71) to the simplified potential vorticity $f(\partial T/\partial z)$ in (4.46) is then apparent.

The pressure-corrected density that can be defined if $\alpha_T = 0$, $\sigma(\rho, p)$ in (9.61) is an example of a potential density (Section 2.5). In general, various potential densities $\sigma_*$ can be constructed by using the equation of state to compute the density, or density anomaly, that a fluid parcel with given potential temperature $\theta$ and salinity $S$ would have at a given reference pressure $p = p_*$:

$$\sigma_* = \tilde{\mathcal{R}}(p_*, \theta, S) \quad \text{or} \quad \sigma_* = \tilde{\mathcal{R}}(p_*, \theta, S) - \rho_0. \qquad (9.73)$$

Potential densities so constructed are materially conserved when $D\theta/Dt = DS/Dt = 0$. However, in the general case, they are functions of both $\theta$ and $S$ and, consequently, are not suitable candidates for the desired function $\Lambda(\rho, p)$ and cannot be used to construct materially conserved potential vorticities.

The simplified equation of state (9.52) also allows the direct demonstration of a curious aspect of the physical properties of seawater. Suppose two equal volumes of seawater, with potential temperatures $\theta_1$ and $\theta_2$ and salinities $S_1$ and $S_2$, respectively, are mixed at constant pressure $p$. Then, the difference between the average of the corresponding specific volumes $1/\rho$ before mixing, $1/\rho_1 = \mathcal{V}(p, \theta_1, S_1)/\rho_0$ and $1/\rho_2 = \mathcal{V}(p, \theta_2, S_2)/\rho_0$, will be different from the specific volume $1/\rho_m$ of the mixture. For, with the temperature $\theta_m$ and salinity $S_m$ of the mixture given by the respective averages

$$\theta_m = \frac{1}{2}(\theta_1 + \theta_2), \quad S_m = \frac{1}{2}(S_1 + S_2), \tag{9.74}$$

the difference in specific volumes, normalized by $1/\rho_0$, is

$$\Delta_m \left( \frac{\rho_0}{\rho} \right) = \frac{\rho_0}{\rho_m} - \frac{1}{2} \left( \frac{\rho_0}{\rho_1} + \frac{\rho_0}{\rho_2} \right)$$

$$= \mathcal{V}(p, \theta_m, S_m) - \frac{1}{2}[\mathcal{V}(p, \theta_1, S_1) + \mathcal{V}(p, \theta_2, S_2)]$$

$$= -\frac{1}{4}\alpha_2(\theta_1 - \theta_2)^2 < 0. \tag{9.75}$$

Thus the density of the mixture will be greater than the average of the densities before mixing, and the mixed fluid will tend to sink. This phenomenon is known as *cabbeling*.

## 9.8 Notes

Additional material on thermodynamics of seawater can be found in the work of Fofonoff (1962). The use of pressure coordinates is standard in meteorology, for which the density variation is of the same order as the reference density (see, e.g., Holton, 1992). The relation between Boussinesq and non-Boussinesq hydrostatic equations is discussed by de Szoeke and Samelson (2004). The simplified nonlinear equation of state (9.52) is taken from the work of de Szoeke (2004).

# 10

# Theory and Observation

## 10.1 A Perturbation-Theoretical Perspective

The basic point of view around which the development in this text has been organized is that of perturbation theory. In general, perturbation theory allows the replacement of a difficult or intractable problem with a simpler approximation that includes the leading-order terms in an expansion of the original variables in powers of a suitable small parameter. In the present case, a formal expansion was not necessary because the effective leading-order terms for large-scale ocean circulation dynamics could be identified by direct scaling of terms in the fundamental equations. These leading-order terms—the perturbation theory for the large-scale ocean circulation—are the planetary geostrophic equations (2.102)–(2.107).

The planetary geostrophic theories of large-scale circulation described in this text provide deep, quantitative insight into the large-scale physical structure of the ocean. For example, the reduced-gravity and ventilated thermocline models (Chapter 5) give persuasive explanations for the basic structure of the upper main subtropical thermocline, including the characteristic downward and westward slope of thermocline isosurfaces (Figure 1.7). Similarly, the Sverdrup interior solution (3.36) for the depth-integrated wind-driven transport may be favorably compared with interior geostrophic transports determined from hydrographic measurements.

Given a solution of an approximate problem constructed using perturbation theory, one would like to know how accurately it represents the solution of the original problem. If the original problem cannot be solved theoretically by other means, as is the case with large-scale ocean circulation, then the comparison must be made to observations of the system under consideration. In principle, such a comparison should result in an objective, quantitative assessment of the accuracy of the perturbation theory.

Despite the successes of planetary geostrophic theory, it is also true that a comprehensive, quantitative assessment of the accuracy of this perturbation theory for the large-scale ocean circulation has not been achieved, half a century after the original derivation of the planetary geostrophic equations. From the point of view of ocean circulation theory, this is a disappointing state of affairs because theoretical models based on these equations provide the primary set of conceptual constructs that inform our physical insight and frame our existing understanding. Although some steps toward such an assessment have been taken using inverse modeling techniques, a generally established, objective measure of the quantitative accuracy of the planetary geostrophic theory is not available.

There are essentially three broad reasons for this persistent gap between theory and observation. First, as noted and demonstrated in previous chapters, the large-scale perturbation theory is degenerate. Second, the major scientific effort on comparison of ocean observations with models has centered on numerical simulations based not on the planetary geostrophic equations but on the primitive equations. Third, the observational record is itself limited, relative to the temporal and spatial scales of interest.

## 10.2 Degeneracy of the Large-Scale Theory

The degeneracy of the large-scale perturbation theory—especially the inability of the planetary geostrophic equations to satisfy a standard set of lateral boundary conditions in closed basins—has been discussed in previous chapters. This degeneracy requires that some representation of the mean effect of small-scale motions on the large-scale flow must be introduced into the large-scale equations to pose consistently all but the most restrictive calculations. The introduction of these representations of small-scale motions inevitably causes a dependence of the resulting solutions on the details of the small-scale model.

The best exceptions to this rule are the Sverdrup interior solution (3.36) for the depth-integrated wind-driven transport, which depends only on the imposed wind stress field and the no-normal-flow condition at the eastern boundary, and the reduced-gravity and ventilated thermocline models of the subtropical upper main thermocline (Chapter 5), which add a dependence on an imposed surface density distribution and an eastern boundary thickness; the latter can perhaps be replaced with some form of gyre-boundary closure, such as (5.40). The simple analytical model of thermal circumpolar flow (Section 7.4), although less well established, may be another such example. These theories are essentially deductive results for limited regimes or elements of the large-scale circulation. Their success depends in substantial part on the happy, independent accident that the Ekman transport, and the critical large-scale boundary condition on vertical velocity, can be obtained directly from the wind stress, without consideration of the detailed structure of the turbulent ocean-surface boundary layer. Such good fortune does not seem to await in the more complex regimes of the large-scale ocean lateral and bottom boundary layers. For example, the western boundary layers of the subtropical gyres have associated nonlinear recirculation regimes and are not purely passive conduits for a boundary-layer return flow required to close the large-scale interior circulation. Similarly, there is evidence that hydraulic effects at bottom topographic constrictions may influence the meridional flow of deep waters from their high-latitude formation regions and thus exert their own control on the dynamics of the large-scale meridional overturning cells.

In general, the formulation of complete models of large-scale circulation based on the planetary geostrophic equations requires the introduction of the supplementary representations of the effects of small-scale motions on the large-scale flow. The approach taken in this text has been to use a minimal set of such models, of maximal simplicity, to emphasize the intrinsic characteristics of the planetary geostrophic flow. Though this approach is adequate to illustrate basic elements of the large-scale circulation, it is difficult to translate into a systematic, quantitative framework that would yield a unique, reliable measure of the accuracy of planetary geostrophic model solutions. Consequently, the degeneracy of the planetary geostrophic perturbation theory has itself been an obstacle to such assessment.

## 10.3 Numerical Simulations of Large-Scale Circulation

In general, analytical or semianalytical solutions of the planetary geostrophic equations for large-scale flow are available in only the most idealized basin geometries. Even in these geometries, important elements of the theory are not closed and describe circulation in only part of a given domain. Similarly, the theory for thermohaline flows, including nonlinearity in the equation of state and sufficiently detailed representation of boundary conditions and small-scale processes to render the temperature and salinity fields dynamically independent, is much less developed than the theory for a single temperature or density variable.

Because of these complexities of the full thermohaline equations, surface boundary processes, and ocean basin geometries, the alternative approach of using numerical methods to solve ocean circulation equations has frequently been taken. Numerical solution methods enjoy certain advantages such as the relatively straightforward inclusion of nonlinear dynamical interactions, of sophisticated schemes for the representation of the effects of small-scale motions on the large-scale flow, and of complex basin geometries and topographies. Consequently, a numerical approach is often preferred when detailed quantitative comparisons with specific observations are to be made or when specific quantitative inferences about past, present, or future ocean circulation are to be drawn. However, they also suffer certain disadvantages, perhaps most notably that conceptual understanding is difficult to extract from individual solutions, relative to analytical functional dependencies. In addition, numerical solutions can be dependent on technical details of the particular numerical methods used for a given simulation and are subject to numerical errors that can be difficult to diagnose and characterize.

For the large-scale ocean circulation, most such numerical simulations are based not on the thermohaline planetary geostrophic equations, as might have been anticipated from a theoretical point of view, but on various versions of the hydrostatic primitive equations, frequently in the Boussinesq form (2.95)-(2.100). The reasons for this are varied. One is a historical accident: the first computer programs that became widely available for large-scale ocean circulation studies were informed by and adapted from computer codes used to simulate the large-scale atmosphere, which were themselves based on a related set of hydrostatic primitive equations for the atmosphere, for which the large first internal deformation radius and the absence of rigid lateral boundaries make the planetary geostrophic regime of less interest. Another is that despite the relative conceptual simplicity of the planetary geostrophic equations, special considerations are necessary to achieve significant computational advantages through their use, and these advantages tend to decrease as the horizontal resolution of the computation is increased, as may be required, for example, to resolve western or equatorial boundary layers. A third is that the planetary geostrophic equations have potential shortcomings as general, time-dependent models, including the equatorial singularity and an ultraviolet catastrophe for baroclinic instability, in which disturbances at arbitrarily small scales grow arbitrarily fast. A fourth is that under many circumstances, the primitive equations should be an accurate approximation for flow on horizontal scales as small as a few kilometers or less; thus a single computer code can be used with progressively finer numerical grids without fundamental changes.

The latter, fourth advantage is, from a theoretical point of view, also a significant disadvantage because it leads to a fundamental dependence of the characteristics of a given primitive-equation simulation on the purely technical choice of numerical grid resolution. Essentially, only a single simplification—the hydrostatic approximation—separates the primitive equations from a much more general set of compressible fluid equations. Consequently, the full range of physical phenomena that can be described by these equations is very wide and includes motions with much smaller horizontal scales and much larger aspect ratios than are represented in numerical solutions in global or basin-scale domains. Numerical convergence can be approached in a primitive equation model if friction and diffusion parameters are fixed at relatively large values as grid resolution is increased, but this convergence is of limited theoretical interest because of the essential dependence of the resulting solution on the parameterizations. Removing this dependence by extending large-scale primitive-equation simulations toward the physical limit of small friction and diffusion requires the numerical resolution of impractically small space and time scales.

Thus primitive equation simulations of large-scale ocean circulation are, in this sense, not well-posed numerical problems: simulations on sufficiently coarse grids capture only planetary geostrophic dynamics, typically with large frictional and diffusive effects, while simulations on progressively finer grids include a broad succession of additional, distinct physical phenomena and mathematical characteristics.

In contrast, there is suggestive evidence that, at least for the case of steady solutions, standard numerical methods can yield approximately convergent results for planetary geostrophic models in the limit of small friction and diffusion. These steady planetary geostrophic solutions may be unstable in that limit but nonetheless are of substantial physical interest as convergent numerical representations of intrinsic, large-scale circulation structures, many of which may be, at least in part, described by analytical theories.

The convergence problem for the primitive equations has complicated the comparison of ocean observations with primitive-equation models, hindering the development of a systematic, objective assessment of the accuracy of those models. In addition, and for similar reasons, the primitive equations have remained stubbornly resistant to theoretical analysis and explicit solution, except under additional asymptotic restrictions: the combination of the range of distinct physical phenomena that they describe and the intrinsic nonlinearity of the equations make them formidable mathematical objects. Consequently, there is little in the way of explicitly primitive-equation theory to which primitive-equation numerical solutions of large-scale ocean circulation can usefully be compared. The absence of such a theoretical framework has been a persistent obstacle in the broad attempt to extract conceptual understanding of large-scale ocean circulation dynamics from primitive-equation simulations.

The main point in the present context, however, is simply that most quantitative evaluation of simulations over the past decades has been carried out for primitive-equation rather than planetary geostrophic models. This focus of effort has in itself limited the possibilities for the development of a well-established, quantitative assessment of the accuracy of the planetary geostrophic description of large-scale circulation.

## 10.4 The Observational Record

The purpose of this text is to give an introduction to the existing theory; a detailed comparison of the theoretical models to the observational record is beyond its intended scope. Particularly for the subtropical gyre and upper main thermocline, existing observations are sufficient to allow relatively detailed qualitative and, to some extent, quantitative comparisons with various elements of the large-scale circulation and thermocline structure. Similarly, syntheses of new and existing observations are yielding progressively refined estimates of essential elements of the meridional overturning circulations and their role in the Earth's climate system. In part to compensate for the difficulty of assessing the accuracy of any circulation theory from purely physical observations, a wide range of methods and techniques, including chemical tracer and other indirect measurements, has been incorporated into such syntheses and comparisons in progressively more sophisticated and ingenious ways. This breadth makes any substantial discussion of such observations and comparisons a complex and subtle endeavor that will not be attempted here.

Historically, the theories of large-scale ocean circulation described in this text have largely not preceded basic observations and descriptions of the large-scale ocean circulation and thermohaline structure. Consequently, the influence of observations on the theory is indisputable. The complexity of the large-scale circulation problem, and the inability to decouple the large-scale dynamics from the small-scale dynamics, except in limited regimes, makes it effectively impossible to proceed theoretically in a purely deductive fashion from fundamental physical principles that have been developed through the traditional method of controlled laboratory experiments. Thus many of the theoretical results described in this text must already be seen in part as descriptions of observed phenomena.

Nonetheless, an enduring problem in our attempts to understand the ocean as a physical system is that ocean physical measurements remain limited in scope and quantity. The number and variety of observations of the global ocean's physical structure and circulation that have now been accumulated, measured against an individual's ability to examine them all one by one, is surely vast. Yet, measured against the intrinsic decadal to centennial time scales of the circulation, the broad range of space and time scales of the full spectrum of ocean physical variability, and the expanse and volume of the global ocean, they remain barely adequate to constrain basic aspects of the large-scale circulation. Perhaps of greatest concern is that the historical record of ocean observations is fundamentally short relative to the intrinsic time scales of the large-scale circulation. Some information on previous ocean conditions can be extracted from the geological record, but this essential limitation of the observational record for large-scale circulation will not soon be removed.

This inescapable fact has the consequence that future advances in our understanding of large-scale ocean circulation, and thus of the ocean's role in the Earth's climate system, must likely rely in large part on theoretical rather than empirical analysis. In turn, our future efforts to predict changing large-scale ocean conditions must similarly rely on that conceptual foundation rather than on a more direct empiricism. Along with its intrinsic scientific interest, these practical considerations argue strongly for continued and increased scientific attention to the theory of the large-scale ocean circulation.

## 10.5 Notes

The many noteworthy numerical simulations of large-scale ocean circulation include, for example, those of Gill and Bryan (1971), Cox and Bryan (1984), Bryan (1987), de Verdiére (1988), Manabe and Stouffer (1988), and Toggweiler and Samuels (1995); this list represents an extremely limited sampling of a large body of work. A useful recent reference on numerical solution methods for mathematical models of large-scale ocean circulation, primarily those based on the hydrostatic, Boussinesq, primitive equations, is the text by Griffies (2004). Comparisons of theoretical results to observations include those by Leetmaa et al. (1977) and Schmitz et al. (1992) for the Sverdrup theory of depth-integrated gyre circulation and by Talley (1985) and de Szoeke (1987) for the ventilated thermocline theory. Keffer (1985) mapped large-scale potential vorticity from ocean observations, and Schmitz (1996a, 1996b) synthesized observations of the large-scale gyre and global meridional overturning circulations. The dynamics of hydraulic flow through bottom topographic constrictions are reviewed by Pratt and Whitehead (2007). Wunsch (1996) and Bennett (2002) provide detailed introductions to modern inverse modeling methods for ocean circulation.

# Exercises

## Exercises for Chapter 1

**Exercise 1.1** Consider the two-dimensional fluid flow represented by the Lagrangian transformation

$$\mathbf{x} = \mathbf{X}(\mathbf{a}, \tau) = (a + \Gamma b \tau, b), \tag{1.1}$$

where $\mathbf{x} = (x, y)$, $\mathbf{a} = (a, b)$, and $\Gamma$ is a constant. For this flow, compute (1) the Lagrangian velocity field $\mathbf{v}(\mathbf{a}, \tau) = \partial \mathbf{X}(\mathbf{a}, \tau)/\partial \tau$, (2) the inverse transformation $\mathbf{a} = \mathbf{A}(\mathbf{x}, t)$ giving the Lagrangian label or initial position of the fluid parcel at Eulerian position $(\mathbf{x}, t)$, (3) the Eulerian velocity field $\mathbf{u}(\mathbf{x}, t) = \mathbf{v}[\mathbf{A}(\mathbf{x}, t), t]$, (4) the divergence $\nabla_h \cdot \mathbf{u}$ of the Eulerian velocity field, and (5) the Eulerian representation of the material derivative operator $D/Dt$. What standard example of fluid flow does this transformation represent? Is this flow compressible or incompressible?

**Exercise 1.2** Consider the evolution in the ocean main thermocline of the concentration $C$ of an injected patch of the inert tracer $SF_6$ (sulfur hexaflouride). Assume that this tracer is materially advected with the fluid motion, except to the extent that it is altered by molecular diffusion, with diffusion coefficient $\kappa_{SF6}$.

1. Derive an evolution equation for the evolution of $\rho C$, the total mass of tracer per unit volume.

2. Use the exact mass conservation equation to derive an evolution equation for the concentration $C$ from the result of step 1.

3. Estimate a value for $\kappa_{SF6}$ based on the known values of the thermal and saline diffusivities, $\kappa_T$ and $\kappa_S$. To which of $\kappa_T$ and $\kappa_S$ should $\kappa_{SF6}$ be more similar? Why?

**Exercise 1.3** 1. Combine the salinity equation and the exact mass conservation equation to obtain an evolution equation for the total mass of salt per unit volume, $\rho S$.

2. For steady flow conditions, for which local rates of change vanish, consider a fluid volume $V_1$ that lies between a connected portion $S_0$ of the sea surface and

171

a continuous surface $\mathcal{S}_1$ of constant salinity $S(x, y, z, t) = S_1$, where the isohaline surface $\mathcal{S}_1$ intersects the sea surface along the boundary of the region $\mathcal{S}_0$. If there is no net evaporation or precipitation so that a no-normal-flow condition applies at the sea surface, show that the integral of the diffusive term over the volume $V_1$ must vanish.

3. Assume that the diffusive term in the equation for $\rho S$ from step 1 has the form of the divergence $\nabla \cdot \mathbf{F}_S$ of a flux $\mathbf{F}_S$. Show that the result of step 2 implies that the integrated salt fluxes into the volume $V_1$ through the surfaces $\mathcal{S}_0$ and $\mathcal{S}_1$ must be equal and opposite.

## Exercises for Chapter 2

**Exercise 2.1** For the main thermocline of the upper ocean, the depth scale $D = 1000$ m is more appropriate than the value of $D$ used in Chapter 2. Similarly, for the deep ocean, the dynamic density deviation scale $\Delta\rho = 0.2 \text{ kg m}^{-3}$ is more appropriate than the value of $\Delta\rho$ used in Chapter 2.

1. For these two modified scale values, estimate corresponding velocity scale values $U$ and $W$ in the two regimes, using appropriate assumptions and balances.

2. For these modified scales, compute the dimensionless ratios used to motivate the approximations in Chapter 2. Determine from the resulting ratios whether the approximations in Chapter 2 remain valid.

**Exercise 2.2** 1. Make a set of graphs, at the discrete depths 0, 1500, 3000, 4500, and 6000 m, that is complementary to Figure 2.2, showing exact and approximate potential temperatures as continuous functions of in situ temperature $T$ at each depth.

2. Find ocean profiles of temperature and salinity with values that lie substantially outside the standard deviation range in Figure 2.1. Identify the corresponding physical reasons for the large departures from the mean.

**Exercise 2.3** Consider a compressible, inviscid fluid in a pipe. Assume that the motions are adiabatic, that the fluid has uniform entropy, and that variations normal to the axis of the pipe may be neglected. An appropriate set of equations is

$$\frac{\partial \rho}{\partial t} + \frac{\partial (u\rho)}{\partial x} = 0, \tag{2.1}$$

$$\rho \left( \frac{\partial u}{\partial t} + u \frac{\partial u}{\partial x} \right) = -\frac{\partial p}{\partial x}, \tag{2.2}$$

$$\rho = \mathcal{R}(p, \eta), \quad \eta = \eta_0 = \text{constant.} \tag{2.3}$$

Here $\rho$ is density, $u$ is $x$ velocity, $p$ is pressure, $\eta$ is entropy, and $\mathcal{R}$ is a specified function giving the equation of state.

1. Identify these three equations by name or physical principle.

2. Linearize these equations about $\rho = \rho_0$, $p = p_0$, $u = 0$, where $\rho_0$ and $p_0$ are constants, and derive a single equation for the pressure fluctuation; that is, let

$$\rho = \rho_0 + \rho', \quad p = p_0 + p', \quad u = u', \tag{2.4}$$

and neglect products of primed quantities. Then eliminate $u'$ and $\rho'$ by differentiation and substitution. Use the notation

$$\left. \frac{\partial \mathcal{R}}{\partial p} \right)_{\eta = \eta_0} = \frac{1}{c_s^2}. \tag{2.5}$$

3. Show that the equation for $p'$ has solutions that propagate with phase speed $c_s$; that is; show that

$$p'(x, t) = F(x - c_s t) + G(x + c_s t) \tag{2.6}$$

satisfies the equation for any smooth (differentiable) $F$ and $G$. What kind of waves are these, that is, what physical phenomenon do they describe?

4. Estimate the value of $c_s$ for seawater from Figure 2.1. Recall that $\sigma = \rho - 1000$ and $\sigma_\theta = \rho_\theta - 1000$ for density $\rho$ and potential density $\rho_\theta$ in kg m$^{-3}$ and that potential density is the density that a parcel would have if it were moved adiabatically (i.e., with no change in entropy or salinity) to a reference pressure. Explain your reasoning and compare your estimate with tabulated values of $c_s$.

5. Compute a Boussinesq approximation to the first two equations: assume that $\rho = \rho_0 + \tilde{\rho}$, where $\tilde{\rho} \ll \rho_0$ and $\rho_0$ is a constant, and that $u$ and $\rho$ vary over the same horizontal scale $L$; use the advective time scale to estimate $\partial \rho / \partial t$; and simplify step 1 appropriately. Then, linearize step 2 and derive a single equation for $p$ from the resulting equations.

6. Show that the approximation to the first equation in step 5 is equivalent to letting $c_s^2 \to \infty$ in step 2, that is, that the same equation for the pressure follows in these two cases. (Assume that pressure fluctuations remain finite as $c_s^2 \to \infty$.)

7. Use step 6 to explain qualitatively how the two different approximations of steps 2 and 5 describe what motion is induced in the fluid when a piston is forced into one end of the pipe.

## Exercises for Chapter 3

**Exercise 3.1** Consider the wind-driven circulation in a closed, northern hemisphere, $\beta$ plane ocean basin that is $L_x = 5000$ km in zonal extent and $L_y = 3000$ km in meridional extent, with western and southern boundaries at $x = 0$ and $y = 0$, respectively. Assume that the western boundary layer is confined to a narrow region of width $\delta$

adjacent to $x = 0$ and that the Sverdrup transport balance holds for $0 \approx \delta < x < x_E = L_x$ and for all $y$, that is,

$$\beta_0 \frac{\partial \Psi}{\partial x} = \frac{1}{\rho_0} \left( \frac{\partial \tau_w^y}{\partial x} - \frac{\partial \tau_w^x}{\partial y} \right), \quad 0 \approx \delta < x < x_E, \quad 0 < y < L_y. \tag{3.1}$$

Let the wind stress $(\tau_w^x, \tau_w^y)$ be

$$\tau_w^x = -\tau_0 \cos \left( \frac{\pi y}{L_y} \right), \quad \tau_w^y = 0, \tag{3.2}$$

where $\tau_0 = 0.1$ N m$^{-2}$, and let the Coriolis parameter be given by $f = f_0 + \beta_0(y - y_0)$, where $\beta_0$ is a constant.

1. Let $y = y_0 = L_y/2$ correspond to 30°N, and compute $f_0$ and $\beta_0$, where the latter are taken to be equal to $f$ and $\beta$ at 30°N, respectively.

2. Plot $\tau_w^x$ versus $y$. Suppose that $\tau_w^x$ has been computed from 10 m winds using the neutral drag law

$$\tau_w^x = \rho_a C_d |U_{10}| U_{10}, \tag{3.3}$$

where $\rho_a$ is air density at sea level pressure and temperature, $C_d = 10^{-3}$ is a drag coefficient, and $U_{10}$ is the wind speed at 10 m altitude. Compute and plot $U_{10}$ versus $y$.

3. Let $\Psi = 0$ on the eastern boundary $x = x_E$, and compute $\Psi(x = 0)$ from the Sverdrup balance (3.1). This will be the value of $\Psi$ outside the boundary layer, at the extreme western edge of the interior, in the limit $\delta \to 0$. Sketch the contours of $\Psi$ versus $x$ and $y$ in the domain of validity of (3.1) and label them in units of $10^6$ m$^3$ s$^{-1}$ = 1 Sv, using an appropriate contour increment.

4. Compute and plot the Sverdrup transport $(U_S, V_S)$ versus $y$ at $x = 0$ and $x = x_E$. What are the corresponding maximum zonal and meridional flow speeds if the transport is distributed evenly over a layer of depth (1) 5000 m and (2) 500 m?

5. Compute the vertically integrated Ekman transport and geostrophic Sverdrup transport fields separately, and plot each of them versus $y$ at $x = 0$ and $x = x_E$. What are the corresponding maximum flow speeds if the transports are distributed, respectively, over an Ekman layer of depth 25 m and a geostrophic interior of depth (1) 5000 m and (2) 500 m? At $y = 0$ and $y = 3000$ km, $\Psi = 0$, and there is no net meridional Sverdrup transport. Do the Ekman and geostrophic Sverdrup transports also separately vanish at $y = 0$ and $y = 3000$ km?

6. Compute and plot the Ekman pumping velocity $W_E$,

$$W_E = \frac{\partial U_E}{\partial x} + \frac{\partial V_E}{\partial y}, \tag{3.4}$$

where $(U_E, V_E)$ is the Ekman transport.

7. Suppose that the western boundary layer is a Stommel layer, obtained by adding linear frictional drag $-rU$ and $-rV$ to the depth-integrated momentum equations so that (3.1) becomes

$$r\left(\frac{\partial^2 \Psi}{\partial x^2} + \frac{\partial^2 \Psi}{\partial y^2}\right) + \beta_0 \frac{\partial \Psi}{\partial x} = \frac{1}{\rho_0}\left(\frac{\partial \tau_w^y}{\partial x} - \frac{\partial \tau_w^x}{\partial y}\right). \tag{3.5}$$

Assume that $r$ (and thus also $\delta$) is small enough that the boundary layer can be represented by the Stommel boundary layer solution, with exponential decay in $x$ and parametric dependence on $y$. For $\beta_0$ as earlier, what is the dimensional value of $r$ that gives a boundary layer width scale $\delta = \delta_S = 100$ km? For this value of $r$, plot the meridional velocity of the western boundary current at $x = 0$ (i.e., on the boundary, at the western edge of the boundary layer), assuming that the boundary current transport is distributed evenly over a layer of depth (1) 5000 m and (2) 500 m.

**Exercise 3.2** Consider geostrophic flows at the bottom $\mathbf{u}_b = (u_b, v_b)$ ranging from 0.1 to 5 cm s$^{-1}$, flowing up a linearly sloping bottom with elevation changes of 500 m over horizontal distances of $\Delta x = 10$ to 5000 km.

1. Compute and contour the induced vertical velocity from this flow, as a function of speed $|\mathbf{u}_b|$ and distance $\Delta x$, that arises from the no-normal-flow condition at the bottom.

2. Compare the resulting vertical velocity to a typical value of surface Ekman pumping velocity $W_E$ in the subtropical gyres.

3. For these geostrophic bottom flows, compute and contour the differential pressure per unit length along lines of constant $H$ and the bottom pressure torque as a function of speed $|\mathbf{u}_b|$ and distance $\Delta x$. Compare the magnitude of this bottom pressure torque to typical values of the torque exerted on the sea surface by wind stress (i.e., the curl of the wind stress) in the subtropical gyres.

**Exercise 3.3** 1. Express the angular momentum conservation principle for the heuristic thin-disk model (Section 3.1) of the Sverdrup vorticity relation in terms of the appropriate solid-body moment of inertia. Use volume conservation to derive a vorticity relation in the Sverdrup form from the time derivative of this conservation statement.

2. Compare the wind stress torque exerted on a subtropical ocean gyre to that exerted by a bicyclist on the bottom-bracket axle connecting the cranks to the pedals of a bicycle by (1) finding the area of sea surface 1000 km from the center of the gyre, over which an integrated 0.1 N m$^{-2}$ stress gives a torque around that center that is equal to the torque exerted by the bicyclist, and (2) comparing the total torque for a wind stress field exerted by the wind stress (3.2) over the gyre to that exerted by the bicyclist. Assume that the mass of the bicyclist is 50 kg and that the length of the crank, from pedal to axle, is 20 cm.

## Exercises for Chapter 4

**Exercise 4.1** For constant buoyancy frequency $N_0 = 1$ cph, depth $H_0 = 5000$ m, and latitudes $\{15, 30, 45\}°$N, and for linear planetary geostrophic dynamics about the rest state with $N(z) = N_0$:

1. Compute the barotropic and first baroclinic deformation radii $\lambda_0$ and $\lambda_1$ and the corresponding long-wave phase speeds $c_0$ and $c_1$. How long would it take each of these waves to cross an ocean basin of 5000 km zonal width at $\{15, 30, 45\}°$N?

2. Graph the corresponding dispersion relations $\omega = \Omega(k)$ for zonal wave numbers $-2\pi/L_x < k < 0$, for $L_x = 250$ km, where $\omega$ is frequency. For what wavelengths are these dispersion relations likely to be inaccurate due to the neglect of short-wave dispersion in the planetary geostrophic approximation?

3. Compute and graph the corresponding vertical mode structures $Z_0(z)$ and $Z_1(z)$ using the quadratic approximation to $Z_0$.

**Exercise 4.2** Consider a layered model of ocean stratification with two moving layers ($N = 2$) and a third, deep motionless layer below, with corresponding temperatures $T_1 = 14°$C, $T_2 = 10°$C, $T_3 = 8°$C, and layer thicknesses $h_1 \approx h_2 \approx 350$ m.

1. Compute the reduced gravities $\gamma_2$ and $\gamma_1$, representative numerical values of the Bernoulli functions $B_2$ and $B_1$, and layer 2 potential vorticity $Q_2$.

2. If layer 2 has uniform potential vorticity, with $h_2 = 350$ m at a central latitude $y = y_0 = 35°$N, by approximately how much does $h_2$ vary between $y = y_0 + 1000$ km and $y = y_0 - 1000$ km? Suppose that layer 2 is at rest. What is the zonal geostrophic velocity in layer 1 that would result from such a meridional change in $h_2$?

3. Suppose that the Ekman pumping at $y = y_0$ is $W_E = -1 \times 10^{-6}$ m s$^{-1}$. What is the corresponding vertically integrated geostrophic Sverdrup transport? Suppose that this transport is divided evenly between layers 1 and 2 and that the magnitudes of the geostrophic velocities in layers 1 and 2 are both equal to the numerical value computed at $y = y_0$ for the layer 1 zonal geostrophic velocity in step 2. Compute the horizontal angle between the geostrophic velocities in layers 1 and 2.

**Exercise 4.3** Consider a two-layer fluid that is in geostrophic motion on a $\beta$ plane with Coriolis parameter $f = f_0 + \beta(y - y_0)$, with a rigid lid at $z = 0$ and a flat bottom at $z = -H_0$. Let the fluid in each layer be homogenous, with densities $\rho_1$ and $\rho_2$, respectively, in the upper layer and lower layers and with $\rho_2 - \rho_1 \ll \rho_0$, where $\rho_0$ is a reference or mean density of the layers. Assume that there is a wind stress forcing at the upper surface resulting in an Ekman pumping vertical velocity $W_E$ at $z = 0$. Let the interface between the layers be located at depth $z = -H_1(x, y, t)$.

1. Integrate the continuity equation $\partial u_j/\partial x + \partial v_j/\partial y + \partial w_j/\partial z = 0$ over each layer $j$, $j = \{1, 2\}$ to obtain equations relating the rates of change of the layer

thicknesses $h_1 = H_1$ and $h_2 = H_0 - H_1$ to the horizontal divergence of the volume flux in each layer and, for layer 1, to the Ekman pumping $W_E$.

2. Linearize the equations in step 1 around a state of rest $H_1 = H = $ constant, $u_j = v_j = 0$, $j = \{1, 2\}$, by substituting $H_1(x, y, t) = H + \eta(x, y, t)$, assuming that $|\eta| \ll H$, and dropping quadratic terms in the small quantities $\eta$, $u$, and $v$.

3. For steady flow $(\partial h/\partial t = 0)$, substitute the geostrophic relations into the linearized equations from step 2 to obtain Sverdrup-balance equations in each layer. Is there any Sverdrup transport in the upper layer? Is there any Sverdrup transport in the lower layer? How would the Sverdrup transport be distributed vertically in this steady linear model if the layers were subdivided into a successively greater number of thinner layers with smaller density differences, in an attempt to represent the continuously stratified ocean?

4. Assuming that the pressure at $z = 0$ is $p(z = 0) = p_0(x, y, t)$, integrate the hydrostatic relations to obtain the pressures $p_1(x, y, z, t)$ and $p_2(x, y, z, t)$ in each layer as a function of $p_0$ and the interface depth $H_1(x, y, t) = H + \eta(x, y, t)$.

5. For unsteady flow with $W_E = 0$, substitute the expressions for the pressure from step 4 into the linearized equations from step 2 to obtain the equations governing linear planetary waves in the two layers.

6. Eliminate the terms in the surface pressure $p_0$ from the two equations from step 5 to obtain one equation for the linearized baroclinic mode. What is the wave speed? Graph the wave speed as a function of the fractional upper-layer thickness $H/H_0$ for $0 < H < H_0$. For fixed $H$, what is the limit of the wave speed as $H_0 \to \infty$?

7. Eliminate the time derivative from the two equations from step 5 to obtain a diagnostic equation relating $p_0$ to $h$. For the baroclinic mode from step 6, how are the horizontal velocities in the two layers related? Graph the ratio $|(u_2, v_2)|/|(u_1, v_1)|$ as a function of the fractional upper layer thickness $H/H_0$ for $0 < H < H_0$. For fixed $H$, what is the limit of this ratio as $H_0 \to \infty$?

8. Find a second solution of the equations from step 6 and 7 in which the interface deformation $h$ is identically zero. To what vertical mode does this solution correspond? Does this solution have a finite wave speed? Why or why not?

### Exercises for Chapter 5

**Exercise 5.1** For the dimensional scales as given in Section 2.1, compute and plot the advective-diffusive and internal boundary layer scale estimates of vertical velocity as a function of turbulent heat diffusivity $\kappa_v$ for $10^{-5}$ m$^2$ s$^{-1} < \kappa_v < 10^{-4}$ m$^2$ s$^{-1}$. Compare these values to typical values of the Ekman pumping velocity $W_E$. For the same range of values of $\kappa_v$, compute the corresponding scale-estimate thermocline thicknesses and compare these to typical values of the advective scale $D_a$. On these plots, indicate the ranges of $\kappa_v$ where the scaling would support the validity of (1) the ventilated thermocline theory for the subtropical upper thermocline and (2) the internal boundary layer theory for the subtropical main thermocline.

**Exercise 5.2** Consider a two-layer reduced-gravity ventilated thermocline model in a northern hemisphere basin that is $L_y = 3000$ km in meridional extent and that extends arbitrarily far westward from an eastern boundary at $x = x_E = 0$. Let the central latitude of the basin ($y = L_y/2$) be at 35°N on a $\beta$-plane, and let the Ekman pumping distribution be

$$W_E = W_0 \sin \frac{\pi y}{L_y}, \quad W_0 = -10^{-6} \text{ m s}^{-1}, \tag{5.1}$$

so that the depth-integrated circulation is clockwise and the Sverdrup transport is equatorward. Let layer 2 outcrop at $y = y_2 = L_y/3$ (so near 30°N), and let

$$\rho_2 - \rho_1 = \rho_0 g \alpha_T (T_1 - T_2) = 26.5 - 25.5 = 1 \tag{5.2}$$

$$\rho_3 - \rho_2 = \rho_0 g \alpha_T (T_2 - T_3) = 27.5 - 26.5 = 1 \tag{5.3}$$

in $\sigma$ units ($\sigma = \rho - 1000$ in kg m$^{-3}$); furthermore, suppose that the eastern boundary depth $h_2(x = 0) = H_2(x = 0) = H_{2E} = 500$ m.

1. Compute $h_1$ and $h_2$ along (1) 35°N and (2) $x = -4000$ km.

2. Plot the results of step 1 overlaid on the following mean sections of $\sigma_\theta$ (where $\sigma_\theta = \rho(\theta, S, p_a)$ [kg m$^{-3}$] $-1000$) in the North Atlantic at (1) 36°N and (2) 55°W, respectively:

1.

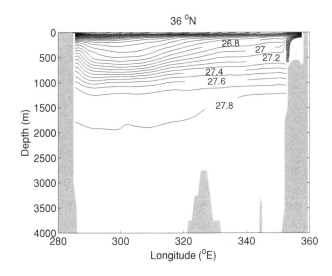

2.

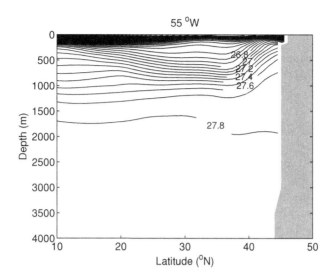

3. Suppose that there is a "climate shift" such that the midlatitude westerlies intensify and the long-term mean Ekman pumping $W_E$ changes from $W_E = W_0(x, y)$ to $W_E = W_1(x, y) = W_0(x, y) + \Delta W(x, y)$, where $W_0 < 0$ and $W_1 < 0$ in the region of interest:

$$\Delta W(x, y) = W_1 - W_0 = \begin{cases} 0, & y \geq y_B, \\ W_2(x, y), & y_A \leq y < y_B, \\ 0, & y < y_A, \end{cases}$$

where $y_B < y_2$. Determine whether, and how, the steady state ventilated thermocline theory predicts that the ocean climate will change in each of the subregions $y \geq y_B$, $y_A \leq y < y_B$, and $y < y_A$ in response to the change in $W_E$.

4. Suppose that $W_E = W_0$ remains fixed but that the outcrop latitude shifts northward from $y = y_2$ to $y = \bar{y}_2 = y_2 + \Delta y_2$, $\Delta y_2 > 0$ ("global warming"). What regions are affected? Compare with step 3.

5. Consider the solution at the subtropical-subpolar gyre boundary $y = y_0$, where $W_E = 0$, layer 1 vanishes, and only layer 2 is in motion. Suppose that the western boundary current is geostrophic so that the geostrophic relation for the meridional velocity holds across the entire basin, and assume that the interface at the base of layer 2 surfaces on the inshore (western) side of the western boundary current at $y = y_0$. Derive an equation relating the total geostrophic transport into the subpolar gyre to a boundary condition for the ventilated thermocline solution.

6. Suppose that $W_E$ and the outcrop latitude remain fixed but that the interhemispheric meridional overturning circulation intensifies so that an increased northward geostrophic flux across the gyre boundary into the subpolar gyre is required in the

layer 2 western boundary current. If the boundary condition appearing in the equation from step 5 adjusts to support this additional flux, what region is affected, according to the ventilated thermocline solution? Compare with steps 3 and 4.

**Exercise 5.3** Consider the planetary geostrophic equations in the form (4.57). Suppose that the surface ($z = 0$) and bottom ($z = -H$) boundary conditions on $w$ and $T$ are

$$w(z = 0) = \frac{\beta}{f^2} W_0, \quad T(z = 0) = -x \frac{T_0}{g\alpha_T}, \tag{5.4}$$

$$w(z = -H) = T(z = -H) = 0, \tag{5.5}$$

where the ocean lies in $x < 0$. Suppose also that $\kappa$ has the form $\kappa = \beta f^{-2} \mu$, where $\mu$ is a constant.

1. Assume that $M$ has the form $M(x, y, z) = xW(z)$, and derive the nonlinear ordinary differential equation and boundary conditions that $W$ must satisfy for this $M$ to solve Equation (4.57). What physical balance does this simplified equation for $W$ represent?

2. Consider the adiabatic, small diffusion limit $\mu \to 0$, and assume that a diffusive internal boundary layer forms at $z = -h$ in the interior, for some $h < H$, with warm fluid above and cold fluid below. In the equation derived in step 1, assume that $W = 0$ at $z = -h$, and expand $W$ in a Taylor series about $z = -h$. Use a boundary layer scaling to determine how the thickness of the diffusive internal boundary layer scales with $\mu$ for the following two cases: (1) $dW/dz > 0$ at $z = -h$ and (2) $dW/dz = 0$ and $d^2W/dz^2 < 0$ at $z = -h$. On general principles, if the depth $h$ represents the depth of the wind-driven planetary geostrophic motion in the subtropical gyre, how would it be expected to scale with the strength $W_0$ of the wind forcing?

3. In the adiabatic, small diffusion limit $\mu \to 0$, show that in the adiabatic regions above and below the diffusive boundary layer, the solution must be quadratic in $z$:

$$W(z) = az^2 + bz + c, \quad -h < z < 0 \tag{5.6}$$

$$W(z) = a'(z + H)^2 + b'(z + H) + c', \quad -H < z < -h. \tag{5.7}$$

Use the boundary conditions, including the assumption $W = 0$ at $z = -h$, and the continuity of $W$ and $dW/dz$ at $z = -h$ to determine $a$, $a'$, $b$, $b'$, $c$, $c'$, and $h$. Is the dependence of $h$ on $W_0$ consistent with either of the scalings from step 2?

## Exercises for Chapter 6

**Exercise 6.1** 1. For velocity and length scales $U$ and $L$ that are appropriate for laminar laboratory rotating-fluid experiments, and for typical values of the molecular diffusivity $\kappa_C$ for dissolved materials or dyes, compute the Peclét number

$Pe_C = UL/\kappa_C$. Conclude whether the homogenization regime can be reached in simple laboratory experiments.

2. Repeat the calculation in step 1 for large-scale recirculating oceanic motions. Use appropriate values of the velocity and length scales $U$ and $L$, and assume values of 100 to 5000 $m^2\ s^{-1}$ for the effective lateral diffusivity from mesoscale eddy motions. Values in this range have been derived as representative of ocean drifter or float dispersion in various mesoscale flow regimes. Conclude whether the homogenization regime can be reached in these oceanic flows. Compare the resulting effective Peclét number to the laminar laboratory values obtained in step 1.

**Exercise 6.2** 1. For $h_{10} = h_{20} = 500$ m, reduced gravities $\gamma_1 = \gamma_2 = 10^{-3}$, and mid-latitude values of $f_0$ and $\beta_0$, compute $W_{0crit}$ in (6.29). Compare the resulting value of $W_{0crit}$ to typical midlatitude values of the Ekman pumping velocity $W_E$.

2. Repeat the comparison in step 1 for a suitable range of values of $h_{10}$, $h_{20}$, $\gamma_1$, and $\gamma_2$, and discuss the implications of the comparisons for the possible relevance of the theory to large-scale ocean flows.

3. For a suitably chosen set of parameter values supporting a recirculation in the subsurface layer, compute and plot versus latitude $y$ the dimensional layer-thickness deviations at $x = 0$ associated with the homogenized and frictionally driven solutions (6.38) and (6.49), respectively.

4. Compute and plot the layer-thickness deviations for a comparable meridional cross section through a homogenized potential-vorticity model (5.66)–(5.67) of the western pool region in the ventilated thermocline theory. Compare this profile to the layer-thickness deviations obtained in step 3.

**Exercise 6.3** Consider a three-layer quasi-geostrophic model with a motionless fourth layer. Suppose that the mean layer depths and density differences are all equal. For large-scale motion, with small dissipation and eddy fluxes, the dimensionless quasi-geostrophic potential vorticity equations are

$$J(\psi_1, q_1) = \frac{f_0}{H_1} W_E - \nabla_h \cdot \mathcal{F}_1, \tag{6.1}$$

$$J(\psi_2, q_2) = -\nabla_h \cdot \mathcal{F}_2, \tag{6.2}$$

$$J(\psi_3, q_3) = -\nabla_h \cdot \mathcal{F}_3, \tag{6.3}$$

where $H_1$ is the constant undisturbed or mean thickness of each layer, $f_0$ is the value of the Coriolis parameter at the central latitude $y = 0$, layer 1 is the surface layer, and $\psi_j$ is the stream function in layer $j$, $j = \{1, 2, 3\}$. The corresponding potential vorticities $q_j$, $j = \{1, 2, 3\}$, are

$$q_1 = \beta y - F(\psi_1 - \psi_2), \tag{6.4}$$

$$q_2 = \beta y - F(\psi_2 - \psi_1) - F(\psi_2 - \psi_3), \tag{6.5}$$

$$q_3 = \beta y - F(\psi_3 - \psi_2), \tag{6.6}$$

where $\beta = \beta_* L^2 / U$ and $F = L^2 / L_R^2$ are dimensionless parameters. Here $\beta_*$ is the dimensional meridional gradient of the Coriolis parameter, $L_R^2 = \gamma H_1 / f_0^2$ is an internal deformation radius, $\gamma$ is the reduced gravity based on the density difference between two adjacent layers, and $U$ and $L$ are the speed and length by which horizontal velocities and distances are scaled, respectively.

Suppose also that the eddy fluxes $\mathcal{F}_j$, $j = \{1, 2, 3\}$ are given by

$$\mathcal{F}_1 = R\nabla_h(\psi_1 - \psi_2), \tag{6.7}$$

$$\mathcal{F}_2 = R\nabla_h(\psi_2 - \psi_1) + R\nabla(\psi_2 - \psi_3), \tag{6.8}$$

$$\mathcal{F}_3 = R\nabla_h(\psi_3 - \psi_2), \tag{6.9}$$

where $R \ll 1$ is a constant, and that the Ekman pumping distribution is given by

$$W_E(x, y) = \begin{cases} -\alpha x, & r = (x^2 + y^2)^{1/2} \le r_1, \\ 0, & r > r_1, \end{cases} \tag{6.10}$$

where $\alpha = W_0 / r_1$, $W_0 > 0$, and $r_1 > 0$ are constants. Assume that $x_W < -r_1 < r_1 < x_E$, where $x_W$ and $x_E$ are the western and eastern ocean boundaries.

1. Solve (6.1)–(6.9) exactly for the barotropic stream function $\psi_B$, where

$$\psi_B(x, y) = \frac{1}{3}(\psi_1 + \psi_2 + \psi_3), \tag{6.11}$$

obtaining an explicit expression for $\psi_B$. Use the boundary condition $\psi_B(x_E, y) = 0$.

2. Show that, neglecting a small term of order $R$,

$$\psi_2 = \Psi_2(\hat{q}_2), \tag{6.12}$$

where $\hat{q}_2 = \beta y + a_1 \psi_B$, and evaluate the constant $a_1$.

3. Show that, neglecting a small term of order $R$,

$$\psi_3 = \Psi_3(\hat{q}_3), \tag{6.13}$$

where $\hat{q}_3 = \beta y + a_2 \Psi_2(\hat{q}_2)$, and evaluate the constant $a_2$.

4. Find the condition on $\alpha$ so that there are closed contours of $\hat{q}_2$.

5. Integrate (6.2) over the area enclosed by a closed isoline of $\hat{q}_2$, and use the results of steps 1 and 2 to solve for $\psi_2$ in terms of $\psi_B$.

6. Using the result of step 5, compute $\hat{q}_3$. How does the region of closed isolines of $\hat{q}_3$, compare with the region of closed isolines of $\hat{q}_2$?

7. Integrate (6.3) over the area enclosed by a closed isoline of $\hat{q}_3$ and solve for $\psi_3$ in terms of $\psi_B$.

8. Solve for the potential vorticities $q_2$ and $q_3$.

9. For forcing strong enough that layer 3 is in motion, sketch $\psi_1$, $\psi_2$, $\psi_3$, $q_1$, $q_2$, and $q_3$.

## Exercises for Chapter 7

**Exercise 7.1** 1. For representative values of the geometric and thermal parameters, compute and contour cross sections of temperature $T$, pressure $p'$, and zonal velocity $u$ for the zonally symmetric analytical circumpolar current (7.26)–(7.28).

2. For a suitably chosen range of representative values of the geometric and thermal parameters, compute the total zonal transport (7.29) of the zonally symmetric analytical circumpolar current (7.26)–(7.28). Compare these values to observed estimates.

**Exercise 7.2** For a value of the meridional slope $s$ in (7.30) estimated from the observed geometry of the Antarctic Circumpolar Current, and other suitably chosen parameter values, compute the Ekman pumping (7.35) and associated Ekman transport required to balance the southward trend of the circumpolar current core. Discuss whether this computed Ekman flow may be consistent with observed wind stress fields at the latitudes of the Antarctic Circumpolar Current.

**Exercise 7.3** Consider the flow of a homogeneous fluid on a $\beta$ plane of fluid in a southern hemisphere basin with the circumpolar gap geometry, a rigid lid at $z = 0$, a flat bottom at $z = -H_0$, and an infinitesimally thin island barrier that spans the gap at the zonal center of the basin. Let the basin boundaries be at $y = \{y_S, y_N\}$ and $x = \{x_W, x_E\}$ and the circumpolar gap cover $y_- < y < y_+$. Let the island barrier be the line segment connecting $(x_0, y_A)$ and $(x_0, y_B)$, where $x_0 = (x_W + x_E)/2$ and $y_S < y_A < y_- < y_+ < y_B < y_N$, so that the tips of the island are outside the latitudes of the circumpolar gap. Suppose that there is no surface forcing, except at the extreme northwest and southwest corners of the basin, where a source and sink, respectively, of equal volume flux magnitude $V$ are located. Assume that the interior flow is described by large-scale, inviscid, planetary geostrophic dynamics and that a meridionally flowing boundary current may occur as required at any western boundary, including the internal western boundary along the east side of the island barrier, but also that no eastern boundary currents occur.

1. Show that the large-scale vorticity dynamics require that any geostrophic flow in the interior be zonal and that the associated pressure gradients be meridional.

2. Suppose that a narrow zonal geostrophic jet with meridional width $\delta \ll y_N - y_S$ exists in the interior, between two regions of different constant pressure, $p_1$ and $p_2$. Integrate the appropriate geostrophic relation meridionally and vertically to obtain a relation between the pressure difference $p_1 - p_2$ and the zonal transport of the jet.

3. Suppose that a boundary current exists along the east side of the island barrier. In which direction must the current extend at the southern or northern tip of the island?

4. Provide an argument showing that because pressure is continuous through the gap, flow from the source to the sink is possible if and only if there is a circumpolar

current through the gap. Sketch the associated flow pattern, which will consist of a system of meridional boundary currents and zonal jets.

5. For the flow pattern sketched in step 4, use the geostrophic transport relations from step 2, and volume transport balances at the junctions of the boundary currents and jets, to develop and solve a system of linear equations determining the interior pressure differences and zonal jet transports in terms of the magnitude $V$ of the volume source and sink.

## Exercises for Chapter 8

**Exercise 8.1** For the analytical model of the warm-water branch of meridional overturning,

1. Estimate $M_E$, $h_m^2$, and $L_{acc}$ from observations. Suppose $M_e = -(1/2)M_E$, and compute the value of $M_e$ Is this value of $M_e$ consistent with observed estimates of North Atlantic Deep Water formation rates and mid-depth meridional overturning volume transport? Explain how the answer depends on assumptions regarding mid-latitude diabatic mixing, that is, on $A_W$.

2. For the values of $h_m^2$ and $M_E$ from step 1, compute $A_e$ and the characteristic eddy flux velocity $v_* = -A_e h_m$. Is this value potentially representative as a velocity scale associated with meridional heat transport by mesoscale eddies?

**Exercise 8.2** For the steady analytical solution (8.27),

1. Compute and contour the value of the eddy flux parameter $A_e$, giving complete eddy compensation ($M_E + M_e = 0$) for suitable ranges of values of the northern hemisphere cooling and southern hemisphere wind parameters, $\delta h_N^2$ and $\tau_1$, with other parameters held fixed at appropriate values. What range, if any, of these values of $A_e$ is physically reasonable?

2. Compute and contour the solution for the eastern boundary depth $h_E$ as a function of $A_e$ and $A_W$ for fixed values of $\tau_0$ and $\tau_1$ and as a function of $\tau_0$ and $\tau_1$ for fixed values of $A_e$ and $A_W$. Discuss the relative strengths of these dependencies in the context of the relative uncertainties of representative values for the corresponding parameters.

**Exercise 8.3** For appropriate ranges of values of the parameters $A_W$ and $A_e$, and of the solution $h_{Es}$, compute and contour the meridional overturning decay time scale $T_{MOC}$ (8.36). Determine the indicated range of physically reasonable values for $T_{MOC}$. Discuss the implications of this time scale for large-scale ocean circulation and climate modeling and prediction.

## Exercises for Chapter 9

**Exercise 9.1** 1. From a suitable climatological atlas or data set, sketch the large-scale distribution of freshwater forcing from the difference of precipitation and evaporation.

Compare the amplitude of the associated vertical velocity to midlatitude values of the Ekman pumping velocity $W_E$.

2. Sketch the depth-integrated Sverdrup flow that is driven by this freshwater forcing. Does it generally oppose or supplement the wind-driven Sverdrup flow?

3. Compute and sketch the effective salt flux that would be used to represent this freshwater forcing in a large-scale, rigid lid, Boussinesq model. Does this flux drive any Sverdrup flow?

**Exercise 9.2** 1. Explore numerical solutions of the thermohaline exchange-flow model (9.12)–(9.14) for a range of values of the free parameters. Find examples of solutions in single- and multiple-equilibrium parameter regimes.

2. Estimate the plausible range of physically representative parameter values for (9.12)–(9.14) when the equations are interpreted as a model of the mid-depth meridional overturning circulation. Does this range include a multiple-equilibrium regime?

**Exercise 9.3** Use the exact spherical differential operators to compute the transformation, analogous to (9.33), of the mass conservation equation for hydrostatically balanced flow into the pressure vertical-coordinate system. Show that for the spherical geometry, the flow is not exactly incompressible in the pressure coordinate system. Identify the essential physical distinction between the pressure and distance radial coordinates that causes this difference and why this distinction becomes unimportant for sufficiently thin fluid layers at sufficiently large radii.

# References

Batchelor, G. K. 1956. Steady laminar flow with closed streamlines at large Reynolds number. *J. Fluid Mech.*, **1**, 177–190.

Batchelor, G. K. 1967. *An Introduction to Fluid Dynamics*. New York: Cambridge University Press.

Bennett, A. 2002. *Inverse Modeling of the Ocean and Atmosphere*. New York: Cambridge University Press.

Bryan, F. O. 1987. Parameter sensitivity of primitive equation ocean general circulation models. *J. Phys. Oceanogr.*, **17**, 970–985.

Burger, A. P. 1958. Scale consideration of planetary motions of the atmosphere. *Tellus*, **10**, 195–205.

Cao, C., Titi, E. S., and Ziane, M. 2004. A "horizontal" hyper-diffusion three-dimensional thermocline planetary geostrophic model: Well-posedness and long-time behaviour. *Nonlinearity*, **17**, 1749–1776.

Cox, M., and Bryan, K. 1984. A numerical model of the ventilated thermocline. *J. Phys. Oceanogr.*, **14**, 674–687.

de Szoeke, R. A. 1987. On the wind-driven circulation of the South Pacific Ocean. *J. Phys. Oceanogr.*, **17**, 613–630.

de Szoeke, R. A. 1995. A model of wind- and buoyancy-forced ocean circulation. *J. Phys. Oceanogr.*, **25**, 918–941.

de Szoeke, R. A. 2004. An effect of the thermobaric nonlinearity of the equation of state: A mechanism for sustaining solitary Rossby waves. *J. Phys. Oceanogr.*, **34**, 2042–2056.

de Szoeke, R. A., and Samelson, R. M. 2004. The duality between the Boussinesq and non-Boussinesq hydrostatic equations of motion. *J. Phys. Oceanogr.*, **32**, 2194–2203.

de Verdiére, A. Colin. 1988. Buoyancy driven planetary flows. *J. Mar. Res.*, **46**, 215–265.

Dewar, W. K., Samelson, R. M., and Vallis, G. K. 2005. The ventilated pool: A model of subtropical mode water. *J. Phys. Oceanogr.*, **35**, 137–150.

Fofonoff, N. P. 1962. Physical properties of sea water. In *The Sea*, vol. 1., edited by M. N. Hill. New York: Interscience, 3–30.

Gill, A. 1968. A linear model of the Antarctic circumpolar current. *J. Fluid Mech.*, **32**, 465–488.

Gill, A. 1982. *Atmosphere-Ocean Dynamics*. New York: Academic Press.

Gill, A., and Bryan, K. 1971. Effects of geometry on the circulation of a three-dimensional southern hemisphere ocean model. *Deep Sea Res. Oceanogr. Abstr.*, **18**, 685–721.

Gnanadesikan, A. 1999. A simple predictive model for the structure of the oceanic pycnocline. *Science*, **283**, 2077–2079.

Griffies, S. 2004. *Fundamentals of Ocean Climate Models*. Princeton, NJ: Princeton University Press.

Holton, J. 1992. *Dynamical Meteorology*. New York: Academic Press.

Huang, R. X. 1988. On boundary value problems of the ideal-fluid thermocline. *J. Phys. Oceanogr.*, **18**, 619–641.

Huang, R. X. 2009. *Ocean Circulation: Wind-Driven and Thermohaline Processes*. Cambridge: Cambridge University Press.

Huang, R. X., and Pedlosky, J. 1999. Climate variability inferred from a layered model of the ventilated thermocline. *J. Phys. Oceanogr.*, **29**, 779–790.

IOC, SCOR, and IAPSO, 2010. The international thermodynamic equation of seawater—2010: Calculation and use of thermodynamic properties. Intergovernmental Oceanographic Commission, Manuals and Guides No. 56, UNESCO (English), 196 pp.

Johnson, H. L., Marshall, D. P., and Sproson, D. A. J. 2007. Reconciling theories of a mechanically-driven meridional overturning circulation with thermohaline forcing and multiple equilibria. *Clim. Dyn.*, **29**, 821–836.

Kamenkovich, V. 1960. The influence of bottom relief on the Antarctic Circumpolar Current. (Translated from Russian.) *Dokl. Akad. Nauk SSSR,* **134**, 983–984.

Keffer, T. 1985. The ventilation of the world's oceans: Maps of the potential vorticity field. *J. Phys. Oceanogr.*, **15**, 509–523.

Leetmaa, A., Niiler, P. P., and Stommel, H. 1977. Does the Sverdrup relation account for the mid-Atlantic circulation? *J. Nav. Res.*, **35**, 1–10.

Ledwell, J. R., Watson, A. J., and Law, C. S. 1998. Mixing of a tracer in the pycnocline. *J. Geophys. Res.*, **103**(C10), 21,499–21,529.

Luyten, J., Pedlosky, J., and Stommel, H. 1983. The ventilated thermocline. *J. Phys. Oceanogr.*, **13**, 292–309.

Manabe, S., and Stouffer, R. J. 1988. Two stable equilibria of a coupled ocean-atmosphere model. *J. Clim.*, **1**, 841–866.

Martinez, C. C., Alanís, E. E., and Romero, G. G. 2002. Determinación interferométrica del coeficiente de difusión de NaCl-$H_2O$, a distintas concentrationes y temperaturas. *Energ. Renovables Medio Ambiente*, **10**, 1–8.

Munk, W. H. 1950. On the wind-driven ocean circulation. *J. Meteorol.*, **7**, 79–93.

Parsons, A. T. 1969. A two-layer model of Gulf Stream separation. *J. Fluid Mech.*, **39**, 511–528.

Pedlosky, J. 1987. *Geophysical Fluid Dynamics*. New York: Springer.

Pedlosky, J. 1996. *Ocean Circulation Theory*. New York: Springer.

Pedlosky, J., and Young, W. 1983. Ventilation, potential-vorticity homogenization and the structure of the ocean circulation. *J. Phys. Oceanogr.*, **13**, 2020–2037.

Prandtl, L. 1905. Uber Flüssigkeitsbewegung bei sehr kleiner Reibung. In *Verhandlungen des dritten Internationalen Mathematiker-Kongresses*, *Heldelberg*, pp. 484–491, Leipzig, Germany: Teubner.

Pratt, L. J., and Whitehead, J. A. 2007. *Rotating Hydraulics: Nonlinear Topographic Effects in the Ocean and Atmosphere*. New York: Springer.

Rhines, P. B., and Young, W. R. 1982a. Homogenization of potential vorticity in planetary gyres. *J. Fluid Mech.*, **122**, 347–367.

Rhines, P. B., and Young, W. R. 1982b. A theory of wind-driven circulation: I. Mid-ocean gyres. *J. Mar. Res.*, **40**(Suppl.), 559–596.

Robinson, A. R., and Stommel, H. 1959. The oceanic thermocline and the associated thermohaline circulation. *Tellus*, **11**, 295–308.

Salmon, R. 1990. The thermocline as an "internal boundary layer." *J. Mar. Res.*, **48**, 437–469.

Salmon, R. 1998. *Lectures in Geophysical Fluid Dynamics*. New York: Oxford University Press.

Samelson, R. M. 1999. Geostrophic circulation in a rectangular basin with a circumpolar connection. *J. Phys. Oceanogr.*, **29**, 3175–3184.

Samelson, R. M. 2004. Simple mechanistic models of mid-depth meridional overturning. *J. Phys. Oceanogr.*, **34**, 2096–2103.

Samelson, R. M. 2009. A simple dynamical model of the warm-water branch of the mid-depth meridional overturning cell. *J. Phys. Oceanogr.*, **39**, 1216–1230.

Samelson, R. M. 2011. Time-dependent adjustment in a simple model of the mid-depth meridional overturning cell. *J. Phys. Oceanogr.*, doi: 10.1175/2010JPO4562.1, in press.

Samelson, R., and Vallis, G. K. 1997. Large-scale circulation with small diapycnal diffusivity: the two-thermocline limit. *J. Mar. Res.*, **55**, 1–54.

Samelson, R. M, Temam, R., and Wang, S. 2000. Remarks on the planetary geostrophic model of gyre scale ocean circulation. *Differ. Integr. Equat.*, **13**, 1–14.

Samelson, R. M., and Wiggins, S. 2006. *Lagrangian Transport in Geophysical Jets and Waves: The Dynamical Systems Approach.* New York: Springer.

Schmitz, W. J., Jr. 1996a. On the world ocean circulation: Vol. I. Woods Hole Oceanographic Institution Tech. Rept. WHOI-96-03, 140 pp.

Schmitz, W. J., Jr. 1996b. On the world ocean circulation: Vol. II. Woods Hole Oceanographic Institution Tech. Rep. WHOI-96-08, 237 pp.

Schmitz, W., Jr. Thompson, J., and Luyten, J. 1992. The Sverdrup circulation for the Atlantic along 24 N. *J. Geophys. Res.*, **97**(C5), 7251–7256.

Stommel, H. 1948. The westward intensification of wind-driven ocean currents. *Eos Trans. AGU*, **99**, 202–206.

Stommel, H., and Arons, A. 1960. On the abyssal circulation of the world ocean: I. Stationary planetary flow patterns on a sphere. *Deep Sea Res.*, **6**, 140–154.

Stommel, H., and Webster, J. 1962. Some properties of the thermocline equations in a subtropical gyre. *J. Mar. Res.*, **44**, 695–711.

Sverdrup, H. 1947. Wind-driven currents in a baroclinic ocean; with application to the equatorial currents of the eastern Pacific. *Proc. Natl. Acad. Sci. U.S.A.*, **33**, 318–326.

Talley, L. D. 1985. Ventilation of the subtropical North Pacific: The shallow salinity minimum. *J. Phys. Oceanogr.*, **15**, 633–649.

Toggweiler, J. R., and Samuels, B. 1995. Effect of Drake Passage on the global thermohaline circulation. *Deep Sea Res.*, Part I, **42**, 477–500.

Truesdell, C. 1954. *The Kinematics of Vorticity.* Bloomington: Indiana University Press.

Tziperman, E. 1986. On the role of interior missing and air-sea fluxes in determining the stratification and circulation of the ocean. *J. Phys. Oceanogr.*, **16**, 680–693.

Vallis, G. 2006. *Atmospheric and Ocean Fluid Dynamics.* New York: Cambridge University Press.

Veronis, G. 1973. Model of the world ocean circulation: I, wind-driven, two-layer. *J. Mar. Res.*, **31**, 228–288.

Webb, D. 1993. A simple model of the effect of Kerguelen Plateau on the strength of the Antarctic Circumpolar Current. *Geophys. Astrophys. Fluid Dyn.*, **70**, 57–84.

Welander, P. 1959. An advective model of the ocean thermocline. *Tellus*, **11**, 309–318.

Welander, P. 1971. The thermocline problem. *Phil. Trans. R. Soc. Lond. A.*, **270**, 415–421.

Wunsch, C. 1996. *The Ocean Circulation Inverse Problem.* New York: Cambridge University Press.

Young, W. R., and Ierley, G. 1986. Eastern boundary conditions and weak solutions of the ideal thermocline equations. *J. Phys. Oceanogr.*, **16**, 1884–1900.

# Index